Analysis, Cryptography and Information Science

Series on Computers and Operations Research

Series Editor: Panos M. Pardalos *(University of Florida, USA)*

Published

Analysis, Cryptography and Information Science

Series on Computers and Operations Research

Series Editor: Panos M. Pardalos *(University of Florida, USA)*

Series on Computers and Operations Research Vol. 10

Analysis, Cryptography and Information Science

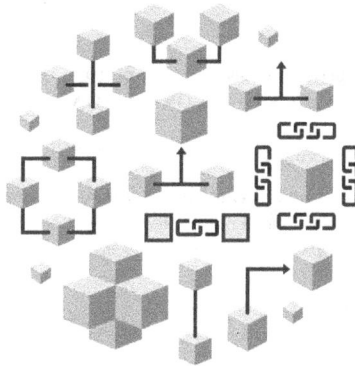

Editors

Nicholas J Daras
Hellenic Military Academy, Greece

Panos M Pardalos
University of Florida, USA

Michael Th Rassias
Hellenic Military Academy, Greece

World Scientific

NEW JERSEY · LONDON · SINGAPORE · BEIJING · SHANGHAI · HONG KONG · TAIPEI · CHENNAI

Published by

World Scientific Publishing Co. Pte. Ltd.

5 Toh Tuck Link, Singapore 596224

USA office: 27 Warren Street, Suite 401-402, Hackensack, NJ 07601

UK office: 57 Shelton Street, Covent Garden, London WC2H 9HE

Library of Congress Control Number: 2023003616

British Library Cataloguing-in-Publication Data
A catalogue record for this book is available from the British Library.

Series on Computers and Operations Research — Vol. 10
ANALYSIS, CRYPTOGRAPHY AND INFORMATION SCIENCE

ISBN 978-981-127-191-5 (hardcover)
ISBN 978-981-127-192-2 (ebook for institutions)
ISBN 978-981-127-193-9 (ebook for individuals)

For any available supplementary material, please visit
https://www.worldscientific.com/worldscibooks/10.1142/13296#t=suppl

Desk Editors: Logeshwaran Arumugam/Steven Patt

Typeset by Stallion Press
Email: enquiries@stallionpress.com

Preface

Analysis, Cryptography and Information Science presents chapters in a broad spectrum of areas of mathematical analysis and its various interconnections to other fields as well as chapters in the domains of cryptography and information science. The chapters of this book also feature the interplay between the above-mentioned domains. Effort has been made for the present work to have a strong interdisciplinary flavor and feature a variety of topics of current vibrant interest and research activity.

The chapters within this book have been contributed by experts from the international community, presenting the essential developments in the corresponding topics as well as problems treated.

Particularly, the contributed chapters discuss topics on one-way actions via holomorphs and split extensions with cryptographic applications, supply chain security by the use of blockchains, cryptographic properties of Boolean functions generating similar De Bruijn sequences, the product subset problem featuring applications to number theory and cryptography, consecutive square-free numbers, cotangent sums related to the Riemann Hypothesis, the approximation of inverse functions of Dirichlet series by rational functions, qualitative queries with fuzzy techniques, topological data analysis and clustering, inequalities for convex functions with applications, as well as general DKH contractions in metric spaces.

We express our warmest thanks to all the contributing authors, who have participated in this collective effort. We would also like to extend our appreciation to the World Scientific staff for their valuable help throughout the publication process of this book.

About the Editors

Nicholas J. Daras is a Professor at the Department of Mathematics and Engineering Sciences at the Hellenic Military Academy. He obtained his Ph.D. in Mathematics with the highest distinction from Université des Sciences et Techniques of Lille Flandres-Artois, Lille, France, in 1988. In 2002, he received a "Best Paper in Mathematics Award" from the Academy of Athens. Over the years, he has supervised 280 diploma theses, 18 postgraduate theses, and one doctoral thesis. He has authored and edited several books. His research interests lie in complex analysis, numerical analysis, modeling and numerical simulation, universal series, holomorphic mappings in several complex variables, rational approximation, operations research, topological quantum computation, numerical representations, as well as quantum cryptography and security.

Panos Pardalos is a Distinguished Emeritus Professor at the Department of Industrial and Systems Engineering at the University of Florida, and an affiliated faculty of the Departments of Biomedical Engineering and Computer & Information Science & Engineering. He is a world-renowned leader in global optimization, mathematical modeling, energy systems, financial applications, and data sciences. He is a Fellow of AAAS, AAIA, AIMBE, EUROPT, and INFORMS and was awarded the 2013 Constantin Caratheodory Prize of the International Society of Global Optimization. In addition, he has been awarded the 2013 EURO Gold Medal prize bestowed by the Association for European Operational Research Societies. This medal is the preeminent European award given to Operations Research (OR) professionals for "scientific contributions that stand the test of time." He has also been awarded the prestigious Humboldt Research Award (2018–2019). The Humboldt Research Award is granted in recognition of a researcher's entire achievements to date — fundamental discoveries, new theories, and insights that have had significant impact on their discipline.

Michael Th. Rassias is an Associate Professor at the Department of Mathematics and Engineering Sciences at the Hellenic Military Academy and a visiting Researcher at the Institute for Advanced Study, Princeton. He obtained his Ph.D. in Mathematics from ETH-Zürich in 2014. During the academic year 2014–2015, he was a postdoctoral researcher at the Department of Mathematics at Princeton University and the Department of Mathematics at ETH-Zürich, conducting research at Princeton. While at Princeton, he prepared with John F. Nash, Jr. (Nobel Prize, 1994 and Abel Prize, 2015) the volume *Open Problems in Mathematics*, Springer, 2016. He has received several awards in mathematical problem-solving competitions, including a Silver medal at the International Mathematical Olympiad of 2003 in Tokyo. He has authored and edited several books, including the edited volume *Analysis at Large* jointly with A. Avila (Fields Medal, 2014) and Y. Sinai (Abel Prize, 2014). His current research interests lie in mathematical analysis, analytic number theory, and more specifically the Riemann Hypothesis, Goldbach's conjecture, the distribution of prime numbers, approximation theory, functional equations, analytic inequalities, and Cryptography.

Contents

Contents

Chapter 1

ℵ-structures: One-way Actions via Holomorphs and Split Extensions with Cryptographic Applications

Iris Anshel*, Dorian Goldfeld†, and Paul E. Gunnells‡

*Veridify Security Inc.,
Shelton, CT, USA*

*ianshel@veridify.com
† dgoldfeld@veridify.com
‡ pgunnells@veridify.com

Let G be a group and $\mathcal{M}$ a mathematical structure with an action of G. The action is said to be *one-way* if given $g \in G$ and $m \in \mathcal{M}$, it is easy to compute $g \cdot m$, but is difficult to recover g given the pair m, $g \cdot m$. In this chapter, we present a general technique to build one-way actions for nonabelian groups that we call ℵ-*structures*. We give examples of ℵ-structures built on the braid group and non-Hopfian groups. As a cryptographic application, we describe a digital signature protocol built on an ℵ-structure.

1.　Introduction

Let G be a group and let $\mathcal{M}$ be an arbitrary mathematical structure. An action of G on $\mathcal{M}$ is a group homomorphism of G into the automorphism group of $\mathcal{M}$. A group action is said to be *one-way* if it can be computed by a polynomial time algorithm but is inherently hard to reverse. In other words, given $g \in G$ and $m \in \mathcal{M}$, the element $g \cdot m$ is easy to compute, but given m and $g \cdot m$, it is difficult to recover g. Group actions that are conjectured to be one-way have been used for cryptographic applications since the inception of public key cryptography. For example, the Diffie–Hellman key exchange [1] and the ElGamal signature algorithm [2] are based on choosing $G = \mathcal{M}$ to be the multiplicative group of a finite field with p elements ($p = $ prime), with the action of modular exponentiation. In later cryptography protocols based on elliptic curves, the group G is the additive group of integers which acts by multiplication on the group of points of an elliptic curve over a finite field (see Ref. 3).

This chapter is focused on actions of nonabelian groups that we conjecture to be one way. In the authors' previous works on a cryptographic hash function [4], a meta public key agreement system [5], and the Walnut digital signature algorithm [6], the authors utilized the action of a representation of the braid group on the direct product of a finite matrix image of a braid group and a symmetric group. The goal of this chapter is to build upon the previous work and describe a unified approach to produce one-way group actions using holomorphs and split extensions. We call this approach an ℵ-*structure* (Definition 1). As an illustration, we describe a new digital signature algorithm built using an ℵ-structure.

With this perspective in place, let G be a group and let $H \leq \mathrm{Aut}(G)$ be a subgroup of the automorphisms of G, acting on the left: for any $h \in H$, $g \in G$, the action of h on g is denoted $^h g$. Recall that the semidirect product $G \rtimes H$ is the group consisting of the ordered pairs

$$\{(g, h) \mid g \in G, h \in H\},$$

where the product of elements (g_1, h_1) and (g_2, h_2) is given by

$$(g_1, h_1) \cdot (g_2, h_2) = (g_1 \cdot {}^{h_1} g_2, h_1 \cdot h_2),$$

and the inverse of the element (g, h) is

$$(g, h)^{-1} = (\,{}^{h^{-1}}(g^{-1}), h^{-1}).$$

Next, let Q be a group and suppose G acts on the *right* as a group of endomorphisms on Q via the mapping

$$\phi: G \longrightarrow \operatorname{End}(Q) \tag{1}$$

as follows. Let $g \in G$ and $q \in Q$. Then $\phi(g) \in \operatorname{End}(Q)$, and we put

$$q^g := \phi(g)(q).$$

With the notation above in place, we have the following definition.

Definition 1 (ℵ-structure). Let G, H, Q, ϕ be as above. An ℵ-*structure*

$$\aleph\big(G \rtimes H, Q, \phi, \star\big)$$

is the data of (i) the groups $G \rtimes H$ and Q, (ii) a homomorphism ϕ of G to $\operatorname{End}(Q)$ giving a right action of G on Q, and (iii) a right action (denoted $\star$) of $G \rtimes H$ on $G \times Q$:

$$(Q \times H) \times (G \rtimes H) \xrightarrow{\ \star\ } (Q \times H),$$

which is given by

$$((q, h_1), (g, h_2)) \longmapsto (q, h_1) \star (g, h_2) \in (Q \times H),$$

where

$$(q, h_1) \star (g, h_2) := \left(q^{\left(^{h_1}g\right)}, h_1 h_2\right).$$

Observe that $\star$ satisfies the following properties:

$$(q, h) \star (1, 1) = (q, h)$$

and

$$((q, h) \star (g_1, h_1)) \star (g_2, h_2) = (q, h) \star ((g_1, h_1) \cdot (g_2, h_2)),$$

and, hence, does define an action. Further, suppose B is a group and that there exists a homomorphism

$$\pi \colon B \longrightarrow G \rtimes H.$$

Then the action $\star$ in the $\aleph$-structure can be extended to B via

$$(q, h) \star b := (q, h) \star \pi(b).$$

If the action of H on G and the function ϕ are difficult to reverse, i.e. are one-way functions, then we expect that the new action $\star$ will likewise be one way.

2.   Examples

2.1. $\aleph$-structure extended to the braid group

Let B_N denote the braid group on N strands, let $\mathbb{F}_q$ denote the finite field of q elements, and for any commutative ring R, let $GL(N, R)$ be the general linear group of invertible $N \times N$ matrices over R. We now view the action given in Ref. 6 as an instance of an extension of an $\aleph(G \rtimes H, Q, \phi, \star)$-structure to the braid group B_N. The relevant groups G, H, Q are the following:

- $G = GL(N, F)$, where F is the ring of Laurent polynomials $\mathbb{F}_q[t_1^{\pm 1}, \ldots, t_N^{\pm 1}]$ in the variables $t_1, \ldots, t_N$.
- $H = S_N$ the symmetric group on N letters, which acts on G by permuting the variables $t_1, \ldots, t_N$.
- $Q = GL(N, \mathbb{F}_q)$.

Finally, we need to define the action $\phi \colon G \to \text{End}(Q)$ which will depend on a fixed $\tau = (\tau_1, \tau_2, \ldots, \tau_N) \in \mathbb{F}_q^N$. Let $g := g(t_1, \ldots, t_N) \in G$. For a fixed set of t-values, $\tau = (\tau_1, \tau_2, \ldots, \tau_N) \in \mathbb{F}_q^N$, we define $g \downarrow_\tau := g(\tau_1, \tau_2, \ldots, \tau_N)$. Then for $m \in Q$, we define the right action

$\phi \colon G \to \mathrm{End}(Q)$ by

$$\phi(g) \ := \ m \ \longmapsto \ m^g = m \cdot g \downarrow_\tau.$$

The above $\aleph\big(G \rtimes H, Q, \phi, \star\big)$-structure can be extended to the braid group B_N via the colored Burau representation (see Ref. 6)

$$\pi \colon B \longrightarrow G \rtimes H.$$

Here, for $\beta \in B_N$, the colored Burau representation is

$$\pi(\beta) = (CB(\beta), \sigma_\beta),$$

where $CB(\beta)$ is the matrix with Laurent polynomial entries in the variables $t_1, \ldots, t_N$ given by the colored Burau representation of β and σ_β denotes the permutation associated with β.

Let

$$(m, \sigma) \in Q \times H = GL(N, \mathbb{F}_q) \times S_N.$$

Then for every $\beta \in B_N$ and fixed $\tau \in \mathbb{F}_q^N$, the action given in Ref. 6 now takes the form

$$(m, \sigma) \star \beta = (m, \sigma) \star \big(CB(\beta),\ \sigma_\beta\big) = (m \cdot (\,{}^\sigma CB(\beta) \downarrow_\tau),\ \sigma \cdot \sigma_\beta).$$

2.2.   ℵ-*structure using the braid inner holomorph*

We now present a second example of an ℵ-structure, which uses the action given in Section 2.1 for fixed $\tau = (\tau_1, \ldots, \tau_N) \in (\mathbb{F}_q^\times)^N$. We denote the action in Section 2.1 by $\star_0$ for clarity.

For $w \in B_N$, the inner automorphism is the function $I_w \colon B_N \to B_N$ given by

$$I_w(\beta) = w\beta w^{-1} \quad (\beta \in B_N).$$

The relevant groups G, Q, H occurring in the $\aleph\big(G \rtimes H, Q, \phi, \star\big)$-structure using the braid inner holomorph $B_N \rtimes \mathrm{Inn}(B_N)$ are defined as follows:

- $G = B_N$,
- $H = \mathrm{Inn}(B_N)$,
- $Q = GL(N, \mathbb{F}_q) \times S_N$.

The right action $\phi\colon G \to \mathrm{End}(Q)$ remains to be defined. Let $\beta \in B_N$. Then for

$$(m, \sigma) \in Q,$$

we define the right action $\phi\colon B_N \to \mathrm{End}\big(GL(N, \mathbb{F}_q) \times S_N\big)$ by

$$\phi(\beta) := (m, \sigma) \longmapsto (m, \sigma)^\beta = (m, \sigma) \star_0 (CB(\beta), \sigma_\beta) \qquad (\beta \in B_N).$$

Let $\big((m, \sigma_0), I_{w_0}\big) \in Q \times H$, i.e. $m \in GL(N, \mathbb{F}_q)$, $\sigma_0 \in S_N$, $w \in B_N$, $I_{w_0} \in \mathrm{Inn}(B_N)$. Then the right action

$$(Q \times H) \times (G \rtimes H) \overset{\star}{\longrightarrow} (Q \times H)$$

is defined by

$$\Big((m, \sigma_0)), \ I_{w_0}\Big) \star (\beta, I_w) = \Big((m, \sigma_0) \star_0 I_{w_0}(\beta), \ I_{w_0 \cdot w}\Big).$$

2.3. $\aleph$-structure using the braid holomorph

Building on Section 2.2, let

- $G = B_N$, $H = \mathrm{Aut}(B_N)$, and $Q = GL(N, \mathbb{F}_q) \times S_N$.

Choosing $\tau = (\tau_1, \tau_2, \ldots, \tau_N) \in B_N$ as before, G acts on Q, again via $\star_0$, and the $\aleph$-structure in this case would utilize the (complete) holomorph of the braid group:

$$B_N \rtimes \mathrm{Aut}(B_N).$$

The right action

$$(Q \times H) \times (G \rtimes H) \overset{\star}{\longrightarrow} (Q \times H)$$

is explicitly given as follows. For

$$\Big((m, \sigma_0), \ \alpha_0\Big) \in \Big(GL(N, \mathbb{F}_q) \times S_N\Big) \times \mathrm{Aut}(B_N)$$

and

$$(\beta, \alpha) \in B_N \rtimes \mathrm{Aut}(B_N),$$

we define the action $\star$ to be

$$\Big((m, \sigma_0), \ \alpha_0\Big) \star (\beta, \alpha) = \Big((m, \sigma_0) \star_0 \alpha_0(\beta), \ \alpha_0 \cdot \alpha\Big).$$

3.   Non-Hopfian Groups and Further Examples

3.1.   *The Baumslag–Solitar group*

With the goal of building more ℵ-structures, we bring the non-Hopfian group into the conversation. A group G is *non-Hopfian* if there is a surjective endomorphism of G *onto* itself

$$\theta : G \longrightarrow G, \quad \theta(G) = G,$$

with nontrivial kernel:

$$\ker(\theta) \ = \ \{g \in G \mid \theta(g) = 1\} \neq 1.$$

In general, given any group G and homomorphism θ whose domain is G, we have the fundamental isomorphism,

$$G/\ker(\theta) \cong \theta(G).$$

Thus, in the case of a non-Hopfian group G, one has the surprising isomorphism

$$G/\ker(\theta) \cong G.$$

Clearly, many familiar groups will *not* have this property and are in fact Hopfian: Hopfian groups include all finite groups, free groups, braid groups, finitely generated linear groups, and more generally, any residually finite group. Since the endomorphism θ has a kernel, it is inherently difficult to reverse and hence serves our purpose.

   One of the early examples of a non-Hopfian group, which dates back to 1962 [7], is the *Baumslag–Solitar group*, denoted by $BS(2, 3)$. The group $BS(2, 3)$ is defined by the presentation

$$BS(2,3) = \langle t, a \mid t^{-1}a^2 t = a^3 \rangle.$$

Let θ denote the endomorphism

$$\theta : t \longmapsto t, \quad a \longmapsto a^2.$$

By construction, the image of θ contains all of $BS(2, 3)$. Indeed, since

$$t, a^2 \in \theta(BS(2,3)),$$

we have

$$t^{-1}a^2 t = a^3 \in \theta(BS(2,3)),$$

and thus, we conclude

$$a = a^3 \cdot a^{-2} \in \theta(BS(2,3)).$$

Since $BS(2,3)$ is generated by $\{a,t\}$, we have

$$\theta(BS(2,3)) = BS(2,3).$$

Given two group elements x, y, let $[x, y]$ denote the commutator $x\,y\,x^{-1}\,y^{-1}$. Then we claim the element

$$[t^{-1}at, a] = (t^{-1}at)\,a(t^{-1}a^{-1}t)a^{-1}$$

is in the kernel of θ. Indeed, we have

$$\begin{aligned}
\theta([t^{-1}at, a]) &= \theta\big((t^{-1}at)\,a(t^{-1}a^{-1}t)a^{-1}\big)\\
&= (t^{-1}a^2t)\,a^2(t^{-1}a^{-2}t)a^{-2}\\
&= a^3\,a^2a^{-3}\,a^{-2} = 1.
\end{aligned}$$

Hence, if one can show $[t^{-1}at, a] \neq 1$, then one can deduce that the kernel $\ker(\theta)$ is nontrivial

$$[t^{-1}at, a] \neq 1 \qquad \Longrightarrow \qquad \ker(\theta) \neq 1,$$

and thus that $BS(2,3)$ is non-Hopfian. The original method [7] of verifying this involves a canonical form for the group $BS(2,3)$, but we remark that one can also show this using a linear representation of $BS(2,3)$.

3.2.   *Additional examples of Non-Hopfian groups*

Before moving to produce ℵ-structures based on non-Hopfian groups, it is valuable to collect more examples from the literature.

Example 1. The group $BS(2,3)$ has a natural generalization [7]: letting p, q denote coprime integers, the group

$$BS(p,q) = \big\langle t, a \mid t^{-1}a^p t = a^q \big\rangle$$

is non-Hopfian. In this example, the surjective endomorphism is given by

$$t \longmapsto t, \quad a \longmapsto a^p,$$

with the nontrivial element in the kernel being

$$[a, t]^p \cdot a^{p-q}.$$

The condition that p, q be coprime can in fact be generalized even further to the case where p, q are *not meshed*; by definition, this means that p, q do not divide each other and their respective sets of prime factors do not coincide.

Example 2. A further extension of the group $BS(p, q)$ (see Ref. 8) above is given by

$$G(\ell, m; k) \;=\; \langle t, a \mid t^{-1} a^{-k} t \, a^\ell \, t^{-1} a^k t = a^m \rangle,$$

where we assume

$$m|k, \; m|\ell, \; \gcd(\ell/m, \, n) = 1.$$

Letting

$$b = t^{-1} a^k t$$

in this example, the surjective endomorphism is given by

$$a \longmapsto a^{\ell/m}, \quad b \longmapsto b, \quad t \longmapsto bt,$$

and the nontrivial element in the kernel is given by

$$a^{-1} b^{-1} a^{-m} \, b \, a \, b^{-1} a^m b.$$

Example 3. An example, where the initial case actually predates those above (see Ref. [9]) is given by

$$\langle x, y, z \mid x^{-1} z x \;=\; y^{-1} z y = z^h \rangle,$$

where we assume $|h| > 1$. In this example, the surjective endomorphism is given by

$$x \longmapsto x, \quad y \longmapsto y, \quad z \longmapsto z^h,$$

and the nontrivial element in the kernel is given by

$$x^{-1} y z y^{-1} x z^{-1}.$$

Example 4. One further example from the literature (see Ref. [10]) is given by

$$\langle a, b, s, t \mid a^{-1}b^{-1}a\,b = 1, s^{-1}as = (a\,b)^2, t^{-1}bt = (a\,b)^2 \rangle.$$

In this example, the surjective endomorphism is given by

$$s \longmapsto s, \quad t \longmapsto t, \quad a \longmapsto a^2, \quad b \longmapsto b^2,$$

and the nontrivial element in the kernel is given by

$$s(a\,b)^{-1}s^{-1}t(a\,b)^{-1}t^{-1}\,s(a\,b)s^{-1}\,t(a\,b)t^{-1}.$$

Example 5. Let F denote the two generator free group,

$$F = \langle s, t \mid - \rangle,$$

and let w denote a nontrivial freely reduced element in F,

$$w \in \langle s, t \mid - \rangle, \quad w \neq 1.$$

We define the group $G(2, 3; w)$ by the presentation

$$G(2, 3; w) = \langle a, s, t \mid wa^2w^{-1} = a^3 \rangle.$$

Let θ_0 denote an automorphism of F that fixes the element w:

$$\theta_0 \colon F \longrightarrow F, \quad \theta_0(w) = w.$$

The automorphism θ_0 can be extended to a homomorphism from $G(2, 3; w)$ to itself,

$$\theta \colon G(2, 3; w) \longrightarrow G(2, 3; w),$$

by specifying the images of the generating set $\{a, s, t\}$,

$$\theta \colon s \longrightarrow \theta_0(s), \quad t \longrightarrow \theta_0(t), \quad a \longrightarrow a^2,$$

and then verifying that θ preserves the defining relator of $G(2, 3; w)$:

$$\begin{aligned}
\theta(wa^2w^{-1}) &= \theta_0(w)\,a^4\,\theta_0(w^{-1}) \\
&= w\,a^4\,w^{-1} \\
&= (w\,a^2\,w^{-1})^2 \\
&= (a^3)^2 \quad \text{(since } wa^2w^{-1} = a^3) \\
&= a^6 = \theta(a^3).
\end{aligned}$$

Observe that θ coincides with θ_0 on the subgroup of $G(2, 3; w)$ generated by $\{s, t\}$. Thus, both generators s, t are in the image of θ, and hence, so is the element $w \in \theta(G(2,3;w))$. To demonstrate that a is in the image of θ, note that since $a^{\pm 2} \in \theta(G(2,3;w))$ and $w \in \theta(G(2,3;w))$,

$$wa^2w^{-1} \in \theta(G(2,3;w)) \implies a^3 = wa^2w^{-1} \in \theta(G(2,3;w)) \implies a$$
$$= a^3 \cdot a^{-2} \in \theta(G(2,3;w)),$$

and we have demonstrated that θ is in fact an endomorphism. Akin to the group $BS(2,3)$ discussed earlier, one can show that this endomorphism also has a nontrivial kernel, and hence,

$$G(2,3;w)/\ker(\theta) \cong G(2,3;w)$$

shows $G(2,3;w)$ to be non-Hopfian. To see this, consider the commutator

$$[waw^{-1},a] = waw^{-1} \cdot a \cdot wa^{-1}w^{-1} \cdot a^{-1}.$$

Assuming for the moment $[waw^{-1},a] \neq 1$, we see that

$$\theta\left([waw^{-1},a]\right) = [wa^2w^{-1},a^2] = [a^3,a^2] = 1,$$

and hence, $1 \neq [waw^{-1},a] \in \ker(\theta)$.

One approach to proving that the commutator $[waw^{-1},a] \neq 1$ requires obtaining a presentation of the subgroup $\langle a \rangle_{G(2,3;w)}$ and then proving that this presentation can be viewed as the fundamental group of a graph of groups, as defined in Bass–Serre [11]. Once this structure is in place, the commutator in question will be an element in a subgroup whose structure is well understood. A second approach to this question is to make the following assumption about the element w: assume that under the projection map

$$F = \langle s,t \mid - \rangle \longrightarrow \langle t \mid - \rangle,$$

defined by $t \mapsto t, s \mapsto 1$ such that

$$w \longmapsto t.$$

With this assumption in place, we see that the analogous projection map

$$G(2,3;w) = \langle a,s,t \mid wa^2w^{-1} = a^3 \rangle \longrightarrow BS(2,3)$$
$$= \langle t,a \mid t^{-1}a^2t = a^3 \rangle$$

is given by

$$[waw^{-1}, a] \longmapsto [tat^{-1}, a] \neq 1,$$

and we deduce $[waw^{-1}, a] \neq 1$.

A more general version to this example is given as follows. Let w, v denote a nontrivial freely reduced element in $F = \langle s, t \mid - \rangle$, and let $G(\{\alpha, \beta, \gamma, \delta\}; w, v)$ be defined by the presentation

$$G(\{\alpha, \beta, \gamma, \delta\}; w, v) = \langle a, s, t \mid waw^{-1} = a^\beta, va^\gamma v^{-1} = a^\delta \rangle.$$

Following the model above, let θ_0 denote an automorphism of F that fixes both w, v, and extend θ to the group $G(\{\alpha, \beta, \gamma, \delta\}; w, v)$ as follows:

$$s \longmapsto \theta(s), \quad t \longmapsto \theta(t), \quad a \longmapsto a^{\alpha\delta}.$$

By construction, we have

$$w\,a^{\alpha\delta}\,w^{-1} = a^{\beta\gamma}\ v\,a^{\gamma\beta}v^{-1} = a^{\beta\gamma} \quad \Longrightarrow \quad v^{-1}\,w\,a^{\alpha\delta}\,w^{-1}\,v = a^{\gamma\beta}.$$

This implies that both $a^{\alpha\delta}$ and $a^{\gamma\beta}$ are in the image of θ and thus the assumption that

$$\alpha\delta - \beta\gamma = 1$$

ensures that θ is a surjection. Akin to the previous example, the element $[v^{-1}waw^{-1}v, a]$ is in the kernel of θ:

$$\theta([v^{-1}w\,a\,w^{-1}v, a]) = \theta(v^{-1}w\,a\,w^{-1}v \cdot a \cdot v^{-1}w\,a^{-1}\,w^{-1}v \cdot a^{-1})$$

$$= \left(v^{-1}w\,a^{\alpha\delta}\,w^{-1}v\right) \cdot a^{\alpha\delta}$$

$$\cdot \left(v^{-1}w\,a^{-\alpha\delta}\,w^{-1}v\right) \cdot a^{-\alpha\delta}$$

$$= a^{\gamma\beta} \cdot a^{\alpha\delta} \cdot a^{-\gamma\beta} \cdot a^{-\alpha\delta} = 1.$$

Proving that the kernel element $[v^{-1}waw^{-1}v, a] \neq 1$ can be approached in various ways is beyond the scope of this chapter. One option is to make the assumption that the subgroup of F generated by w, v,

$$\langle w, v \rangle < F,$$

is freely generated by $\{w, v\}$. By analyzing the normal subgroup generated by $\{a\}$, again using Bass–Serre theory, we can prove the kernel element is indeed nontrivial.

3.3.   ℵ-*Structures based on Non-Hopfian groups*

Having presented various non-Hopfian groups, we turn now to ℵ-structures that can be based on them. In the following cases, the action of the group G on the group Q will utilize a homomorphism

$$\theta \colon G \longrightarrow Q,$$

followed by the right regular representation of Q. Thus, for a fixed $g \in G$, the action of g on an element $q \in Q$ is given by the function

$$\phi(g) := q^g = q \cdot \theta(g).$$

Given a non–Hopfian group G and its associated surjective endomorphism θ whose kernel is nontrivial, we can consider both the inner holomorph and the complete holomorph, respectively:

$$G \rtimes \mathrm{Inn}(G), \quad G \rtimes \mathrm{Aut}(G).$$

Focusing on the inner holomorph, $G \rtimes \mathrm{Inn}(G)$, the ℵ-structure in this case takes the form,

$$\aleph\Big(G \rtimes \mathrm{Inn}(G), \ G \times \mathrm{Inn}(G), \ \phi, \ \star \Big),$$

$$(G \times \mathrm{Inn}(G)) \times (G \rtimes \mathrm{Inn}(G)) \xrightarrow{\ \star\ } (G \times \mathrm{Inn}(G)),$$

where given $g_1, g_2 \in G$ and $I_{x_1}, I_{x_2} \in \mathrm{Inn}(G)$, we have

$$((g_1, I_{x_1}), (g_2, I_{x_2})) \longmapsto (g_1, I_{x_1}) \star (g_2, I_{x_2}) \in (G \times \mathrm{Inn}(G)),$$

which is explicitly given by

$$(g_1, I_{x_1}) \star (g_2, I_{x_2}) = \Big(g_1 \cdot \theta\big(^{I_{x_1}} g_2\big), I_{x_1 \cdot x_2}\Big).$$

Since θ has a nontrivial kernel and hence is difficult to reverse, it leads to an action that likewise is hard to reverse. The case of the complete holomorph similar to the $\aleph$-structure in this case takes the form,

$$\aleph\Big(G \rtimes \mathrm{Aut}(G),\ G \times \mathrm{Aut}(G),\ \phi,\ \star \Big),$$

$$(G \times \mathrm{Aut}(G)) \times (G \rtimes \mathrm{Aut}(G)) \xrightarrow{\ \star\ } (G \times \mathrm{Aut}(G)),$$

where given $g_1, g_2 \in G$ and $\alpha_1, \alpha_1 \in \mathrm{Aut}(G)$, we have

$$((g_1, \alpha_1), (g_2, \alpha_2)) \longmapsto (g_1, \alpha_1) \star (g_2, \alpha_2) \in (G \times \mathrm{Aut}(G)),$$

which is explicitly given by

$$(g_1, \alpha_1) \star (g_2, \alpha_2) \ =\ (g_1 \cdot \theta(\,{}^{\alpha_1}g_2),\ \alpha_1 \cdot \alpha_2).$$

As a last example, we focus on the case of a non-Hopfian group G (with its associated surjective endomorphism θ whose kernel is nontrivial), where there is a natural free image. This is in fact the case for every example considered in this chapter. To complete the conversation on a concrete note, we focus on the case of the group

$$BS(2,3) = \langle t, a \,|\, t^{-1}a^2 t = a^3 \rangle.$$

Let $N \trianglelefteq BS(2,3)$ denote the normal subgroup of $BS(2,3)$ which is the kernel of the surjective homomorphism

$$BS(2,3) \longrightarrow \langle t \,|\, - \rangle.$$

Then N is the normal closure of the generator a, which can be generated by the collection of group elements

$$N \ =\ \big\{ a_k = t^k \, a \, t^{-k} \ |\ k \in \mathbb{Z} \big\},$$

subject to the relations

$$a_{k-1}^2 = a_k^3.$$

Since $\langle t \,|\, - \rangle$ is a free group, the short exact sequence

$$1 \to N \to BS(2,3) \to \langle t \,|\, - \rangle$$

must split, i.e. the group $BS(2,3)$ is the semidirect product

$$BS(2,3) \ =\ N \rtimes \langle t \,|\, - \rangle.$$

By restricting the surjective endomorphism θ, which maps

$$t \longmapsto t, \quad a \longmapsto a^2,$$

we see that

$$\theta \colon N \longrightarrow N,$$

and hence, we have the ℵ-structure

$$\aleph\left(N \rtimes \langle t \mid - \rangle, N \times \langle t \mid - \rangle, \phi, \star\right).$$

The action takes the form

$$\left(N \times \langle t \mid - \rangle\right) \times \left(N \rtimes \langle t \mid - \rangle\right) \xrightarrow{\star} \left(N \times \langle t \mid - \rangle\right),$$

where given $n_1, n_2 \in N$ and $t^{k_1}, t^{k_2} \in \langle t \mid - \rangle$, we have

$$\left((n_1, t^{k_1}), (n_2, t^{k_2})\right) \longmapsto (n_1, t^{k_1}) \star (n_2, t^{k_2}) \in N \times \langle t \mid - \rangle,$$

which is explicitly given by

$$(n_1, t^{k_1}) \star (n_2, t^{k_2}) = \left(n_1 \cdot \theta\left(t^{k_1} g_2 t^{-k_1}\right), t^{k_1+k_2}\right).$$

4. A Digital Signature Based on an ℵ-Structure

Having described a few examples of ℵ-structures, we move to describing an application: a digital signature based on the ℵ-structure. In this application, the ℵ-structure that the signer and verifier(s) will use in the signature method, termed ℵ-DSA, assumes that the action ϕ is a composition of a homomorphism with nontrivial kernel,

$$\theta \colon G \longrightarrow Q,$$

together with the right regular representation of Q onto itself: for $g \in G$ and $q \in Q$,

$$q^g = q \cdot \theta(g).$$

Given an element $(q, h_0) \in Q \times H$, a stabilizer of (q, h_0) is an element $(g, h) \in G \rtimes H$ such that $(q, h_0) \star (g, h) = (q, h_0)$. In our case, since

$$(q, h_0) \star (g, h) = (q \cdot \theta(^{h_0}g), h_0 h),$$

for the element (g, h) to be a stabilizer we see that $h = 1$, and further,

$$\theta(^{h_0}g) = 1.$$

Hence,

$$^{h_0}g \in \text{Ker}(\theta) \implies g \in {}^{h_0^{-1}}\text{Ker}(\theta).$$

Thus, we have $\text{Stab}\,((1,1))$ given by

$$\text{Stab}\,((1,1)) = \text{Ker}(\theta) \times \{1\}.$$

Observe that given $g \in \text{Ker}(\theta)$, the conjugate $(q, h_0)^{-1}(g, 1)(q, h_0)$ is given by

$$(q, h_0)^{-1}(g, 1)(q, h_0) = \left({}^{h_0^{-1}}(q^{-1}gq), 1\right).$$

Since the kernel of θ is a normal subgroup, $q^{-1}gq \in \text{Ker}(\theta)$, we see that

$$(q, h_0)^{-1}(g, 1)(q, h_0) = \left({}^{h_0^{-1}}(q^{-1}g\,q), 1\right) \in \text{Stab}\,((q, h_0)).$$

One very useful feature of stabilizing elements is their inherent ability to allow for an element (g, h) to be obscured without impacting the action of the said element. To be specific, given a product of two elements $(g_1, h_1) \cdot (g_2, h_2) \in G \rtimes H$, if $(g_c, 1)$ stabilizes $(1, 1) \star (g_1, h_1)$, then

$$(1, 1) \star ((g_1, h_1) \cdot (g_2, h_2)) = (1, 1) \star ((g_1, h_1) \cdot (g_c, 1) \cdot (g_2, h_2)).$$

Clearly, the element

$$(g_1, h_1) \cdot (g_c, 1) \cdot (g_2, h_2),$$

will be quite different that the original product $(g_1, h_1) \cdot (g_2, h_2)$, but the action on $(1, 1)$ is the same. Were we to start with a word w of length ℓ in $G \rtimes H$,

$$w \in G \rtimes H,$$

and iterate this insertion process κ times, we would arrive at new element,

$$\kappa(w) \in G \rtimes H,$$

which obscures the original one (particularly when a word rewriting algorithm r is applied to the element $\kappa(w)$, see (2)).

The ℵ-DSA proceeds as follows:

Message Signing

- The signer has a private key

$$\text{priv}(S) = (g_S, h_S) \in G \rtimes H$$

and a public key

$$\text{pub}(S) = (1, 1) \star (g_S, h_S) \in Q \times H.$$

The public key is available to any verifier.
- The signer S initiates the signing of a message M by evaluating the hash of the message $H(M)$, where the hash function is preset and known to both signer and all verifiers.
- The signer S uses a publicly known encoding algorithm that inputs the hash of the message M and produces a unique element

$$E(M) = (g_{H(M)}, 1) \in G \rtimes H.$$

- The signer then applies a group element rewriting algorithm

$$r : G \rtimes H \longrightarrow G \rtimes H, \tag{2}$$

which uses the relations of the group to render a word unrecognizable (examples of such rewriting algorithms can be found in Ref. [6]). Using r, together with the above method of iterative inserting of stabilizing elements κ, the signer produces the signature of the message M: letting

$$\boldsymbol{S}(M) = \boldsymbol{r}\left(\boldsymbol{\kappa}\left((g_S, h_S)^{-1} E(M)(g_S, h_S)\right)\right),$$

the signature of the message M is the ordered pair

$$\text{Sig}(M) = (\boldsymbol{S}(M), M).$$

Signature Verification

- The verifier takes the received signature $\text{Sig}(M)$ and
  - (i) the hash of the message, $H(M)$,
  - (ii) the encoding of $H(M)$, $E(M)$,
  - (iii) the action of the encoded hash of the message $E(M)$ on $(1, 1)$,

$$(1, 1) \star E(M) \in Q \times H.$$

- The verifier then computes

$$\mathrm{pub}(S) \star \boldsymbol{S}(M)$$

and checks that

$$\mathrm{pub}(S) \star \boldsymbol{S}(M) = E(M) \cdot \mathrm{pub}(S) \in Q \times H,$$

where in the above identity, the multiplication is in the direct product $Q \times H$. The signature is verified if the identity holds and rejected otherwise.

5. Concluding Remarks

In this chapter, we chose to present a digital signature algorithm based on the introduced ℵ-structure whose security is directly tied to group actions which are conjectured to be one-way. In future work, we plan to expand the possible cryptographic applications. Further, there are many other classes of groups and group actions that have the potential to be the basis for secure cryptographic primitives. The authors perceive it would be interesting to explore the following groups: Grigorchuk Groups [12], Thompson Groups [13], and Artin–Tits Groups [14].

References

[1] W. Diffie and M. E. Hellman (1976). New directions in cryptography. *IEEE Trans. Inform. Theory*, **IT-22**(6), 644–654. doi: 10.1109/tit.1976.1055638.
[2] T. ElGamal (1985). A public key cryptosystem and a signature scheme based on discrete logarithms. *IEEE Trans. Inform. Theory*, **31**(4), 469–472. doi: 10.1109/TIT.1985.1057074.
[3] D. Hankerson, A. Menezes, and S. Vanstone (2004). *Guide to Elliptic Curve Cryptography*. Springer Professional Computing (Springer-Verlag, New York, 2004). doi: 10.1016/s0012-365x(04)00102-5.
[4] I. Anshel, D. Atkins, D. Goldfeld, and P. E. Gunnells (2016). A class of hash functions based on the algebraic eraser™. *Groups Complex. Cryptol.*, **8**(1), 1–7. doi: 10.1515/gcc-2016-0004.

[5] I. Anshel, D. Atkins, D. Goldfeld, and P. E. Gunnells (2021). Ironwood meta key agreement and authentication protocol. *Adv. Math. Commun.*, **15**(3), 397–413. doi: 10.3934/amc.2020073.

[6] I. Anshel, D. Atkins, D. Goldfeld, and P. E. Gunnells (2020). WalnutDSA^{TM}: A group theoretic digital signature algorithm. *Int. J. Comput. Math. Comput. Syst. Theory*, **6**(4), 260–284. doi: 10.1080/23799927.2020.1831613.

[7] G. Baumslag and D. Solitar, Some two-generator one-relator non-Hopfian groups, *Bull. Amer. Math. Soc.*, **68**, 199–201. doi: 10.1090/S0002-9904-1962-10745-9.

[8] A. M. Brunner (1980). On a class of one-relator groups, *Canadian J. Math.*, **32**(2), 414–420. doi: 10.4153/CJM-1980-032-8.

[9] G. Higman (1951). A finitely related group with an isomorphic proper factor group. *J. London Math. Soc.*, **26**, 59–61. doi: 10.1112/jlms/s1-26.1.59.

[10] D. Meier (1982). Non-Hopfian groups. *J. London Math. Soc. (2)*, **26**(2), 265–270. doi: 10.1112/jlms/s2-26.2.265.

[11] J.-P. Serre (2003). *Trees.* Springer Monographs in Mathematics (Springer-Verlag, Berlin). Translated from the French original by John Stillwell, Corrected 2nd printing of the 1980 English translation.

[12] R. I. Grigorchuk (1984). Degrees of growth of finitely generated groups and the theory of invariant means. *Izv. Akad. Nauk SSSR Ser. Mat.*, **48**(5), 939–985.

[13] J. W. Cannon, W. J. Floyd, and W. R. Parry (1996). Introductory notes on Richard Thompson's groups. *Enseign. Math. (2)*, **42**(3-4), 215–256.

[14] J. Tits (1966). Normalisateurs de tores. I. Groupes de Coxeter étendus. *J. Algebra*, **4**, 96–116. doi: 10.1016/0021-8693(66)90053-6.

Chapter 2

Using Blockchains to Support Supply Chain Security

Ioannis T. Christou*,†,‡‡, Sofoklis Efremidis†,‡,§§, Giannis Klian§,¶¶,
Gerasimos C. Meletiou¶,‖,‖‖, and Michael Th. Rassias**,††,***

*Netcompany-Intrasoft, Peania, Greece
†American College, Agia Paraskevi, Greece
‡Maggioli Greek Branch, Marousi, Greece
§CodeHub, Athens, Greece
¶Department of Agriculture, University of Ioannina, Arta, Greece
‖Computational Intelligence Laboratory (CILab), University of Patras,
Patras, Greece
**Department of Mathematics and Engineering Sciences, Hellenic Military
Academy, Vari Attikis, Greece
††Institute for Advanced Study, 1 Einstein Dr, Princeton, NJ, USA

‡‡ichristou@acg.edu
§§sofoklis.efremidis@maggioli.gr
¶¶iklian@athtech.gr
‖‖gmelet@uoi.gr; gmelet@neptune.math.upatras.gr
***mrassias@sse.gr

This chapter presents a novel approach for enhancing security and trust
in supply chains through the use of blockchain technology. Supply chains
lie at the base of world's economy and typically comprise numerous stakeholders, which share no trust relationships while at the same time they
need to interact and cooperate through complex processes. Cooperation

between stakeholders presumes agreement between them at the different stages of their interaction, something that may be challenging to achieve. This chapter shows how blockchains can be used for logging stakeholder interactions, guaranteeing consensus among them, and implementing complex service agreements that may involve financial transactions through escrow accounts.

1. Introduction

This section gives an overview of the main concepts used in the rest of the chapter. In particular, it presents supply chains, distributed ledgers and blockchains. It then highlights how blockchain technology can be used to support typical operations of supply chains.

1.1. *The complex world of supply chains*

A number of definitions exist for what a supply chain is. According to Ref. [1], *a supply chain is a system of organizations, people, activities, information and resources involved in supplying a product or service to a consumer.* A classical supply chain typically starts with the mining of raw materials and ends with the delivery of finished products to consumers. It involves numerous stakeholders that interact through complex processes, each one receiving components, materials, parts from its suppliers, performing some processing or transformation to them and pushing the result to the next stakeholder down the chain until the finished product or service reaches a consumer. When the products or services that reach the consumers are digital, similar interactions between supply chain stakeholders take place. In this case, the supply chain involves the development and integration of digital components and artifacts instead of mining of raw material.

Supply chains form the backbone of global economy and any disruptions to their operations have profound financial, social and political implications. In the complex ecosystem of supply chains, issues like security and resilience are of prime importance and they both constitute crucial requirements for guaranteeing their smooth operations. In this direction, significant research effort is put toward the formal certification of security and resilience qualities of supply chains. As an example, the EU project CYRENE [2] proposes and specifies a certification scheme and a conformity assessment process for the certification of security requirements for supply chain services, thus contributing toward the European Cybersecurity Act [3]. The CYRENE certification scheme is an extension and specialization of the candidate Cybersecurity Certification Scheme of ENISA [4] to supply chain services.

This chapter addresses issues of security and trust in the complex supply chain ecosystem through the use of blockchain technology. It shows how distributed ledgers and in particular blockchains can be used for enhancing the security and trust between supply chain stakeholders that may not share any trust relationships and how smart contracts can be used for implementing service level agreements (SLAs) between them.

1.2. *Distributed ledgers and blockchains*

A distributed ledger is a consensus-based distributed database in which new data can be added, but previously added data are immutable. This implies that every new datum that is added becomes immutable upon its insertion to the database. The replicas of the ledger are identical and are constructed as a result of a consensus mechanism among all participating entities or stakeholders.

A distributed ledger is the result of agreement between parties that share no trust relationships. Moreover, it guarantees that whatever has been agreed in the past remains immutable. To this end, distributed ledgers provide a solution to a well-known problem in distributed systems, namely, that of the Byzantine generals problem [5].

Blockchains are special kinds of distributed ledgers whereby the database is formulated as a list of data [6]. New data are appended to the list and if they are accepted, they become immutable through cryptographic techniques. Cryptographic techniques are also used for guaranteeing the integrity of the data stored in blockchains and the validity of operations on them. The same list is replicated to all participating nodes of the blockchain network and is the result of consensus among its nodes. Blockchains were introduced in Ref. [7] where it is shown how they can be used for implementing the Bitcoin cryptocurrency. In the same chapter, it is shown how some fundamental problems that arise with cryptocurrencies, like the double-spending problem, are addressed through blockchains.

1.3. *Blockchain support of supply chains*

Blockchains provide a solution to support supply chain services in which participating stakeholders must reach agreement in the presence of failures. Blockchains meet the agreement requirements as data can be immutably logged only as the result of consensus between stakeholders. Logged data pertain to supply chain transactions and may provide at the same time a trace of the flow of material, components, services or other entities from one stakeholder to another. Moreover, SLAs between supply chain stakeholders can naturally be implemented as smart contracts, whose execution can be automated and their results logged into the blockchain itself, thus enhancing the self-governance of the whole supply chain. For example, a payment may be initiated upon acceptance of a product or service by a consumer and all requirements, conditions or clauses that have been agreed to be automatically checked by the smart contract. Upon satisfaction of contractual requirements between stakeholders, processes like transfer of funds may be executed automatically.

1.4. *Structure of the chapter*

This chapter is structured as follows. Section 2 gives an overview of blockchain technologies. Section 3 presents some security issues of blockchains. Section 4 gives an overview of some indicative applications of blockchain technologies. Section 5 gives an overview of Ethereum and Hyperledger. Section 6 presents a prototype implementation of a supply chain simulator that makes use of blockchain technologies. Finally, Section 7 concludes the chapter.

2. Background on Blockchains

Blockchains are immutable time-stamped lists of data records, which are managed by the users of the network. The chain starts with what is called a *genesis* block, which is the first time-stamped record added in the chain. Due to the peer-to-peer distributed architecture of blockchains, new blocks can only be added to the chain if the majority of the network accepts them. This is enabled by a consensus algorithm that helps in reaching agreement on the new datum to be added by the nodes in the network. Each block is secured through a

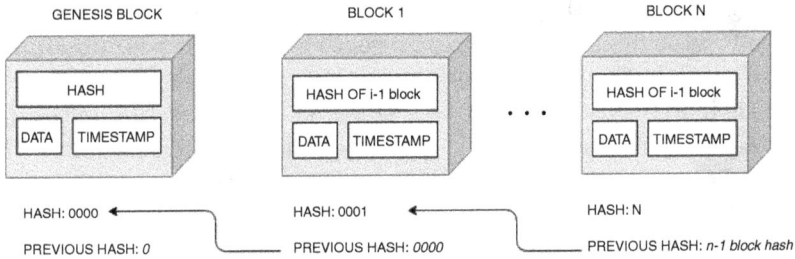

Figure 1. Representation of blockchains.

unique value, which is computed from a cryptographic hash function applied to it. Except for the genesis block, which is given a special hash value, all other blocks contain a reference to the hash of the previous block; this is the mechanism by which the blocks of the chain are securely linked together. The strength of the chain comes from the fact that each block refers to the hash value of the previous block, as shown in Figure 1. This design results in the immutability of the contents of the blockchain.

An attack on a blockchain to change a value stored in a block will result in a different hash value for the whole block, something that will affect the hash value of the next block and these changes will propagate to the last block of the chain. A block is constructed (or mined) by the peers of the blockchain network, a process that is computationally heavy, therefore an attempt to change a value in the middle of the chain will require enormous resources for updating the chain.

Block creation (mining) is the process of securing a set of transactions in a block and the consequent agreement of all peers that participate in a blockchain network. Block mining involves solution to a computationally difficult problem, a process that proves that there is incentive for it and work has been expended so as to safeguard against attacks on the network. There are two approaches to the mining process: the proof of work (PoW) by which effort (in terms of computational power) is expended to secure a block and proof of stake (PoS) by which the more powerful (in terms of merit) nodes of the network (which also have larger incentive for protecting the blockchain) have higher decision power. The Bitcoin blockchain and also the Ethereum blockchain make use of the PoW approach to mining.

In general, several miners compete to solve the same computationally difficult problem, and the one who solves it first announces the solution to the others. Upon reception of a solution, other miners stop their mining efforts, validate the announced solution and if the validation succeeds, they add the block to the end of the chain. This is the mechanism by which consensus between the peer notes of the network is achieved.

The advantages provided by blockchains are anonymity, security, immutability, transparency and lack of any centralized authority. Anonymity is achieved by breaking the link between human identity and its representation in the blockchain, as user identities are represented by addresses which result from cryptographic hash functions. Security and immutability result from the fact that information that is stored on the chain cannot be deleted or altered easily, as explained above. Transparency is enabled by the fact that the same blockchain is replicated to all nodes of a network, so all nodes can inspect all blockchain data. Finally, blockchains are, by their design, distributed and replicated databases whereas consensus for their updates is achieved by (distributed) network peers and not any centralized authority.

Blockchains like Ethereum can accommodate smart contracts, which represent SLAs between participants of the blockchain. Smart contracts can be programmed in programming languages and their execution can be automated in a distributed manner, without any central control, thus enhancing the self-governance of the network. Results that are produced by the execution of smart contracts can be written permanently onto the blockchain.

3. Security Aspects of Blockchains

This section gives an overview of some of the security aspects of blockchains, focusing on those that relate to the Bitcoin blockchain. The Bitcoin blockchain makes use of public key cryptography as well as of cryptographic primitives for maintaining the anonymity of its users, logging transactions on the chain and maintaining the integrity of the chain itself. Bitcoin blockchain users are anonymous and are identified with addresses that are generated through cryptographic transformations on their public key, as presented in the following section. The generation of the public key makes use of elliptic curves, which are briefly presented in the sequel. Finally, transactions that are logged on the chain are signed by the private key of the user who generates them.

3.1. *Keys and address generation*

In public key (asymmetric) cryptography [8], a pair of keys (a private and a public) are used for the secure communication between parties. The two keys are mathematically interrelated and are used for encrypting and decrypting exchanged messages. Private keys are unique and secret to the user who creates them. To send a confidential message to a user, the sender uses the recipient's public key. The message can only be decrypted with the corresponding private key, which is kept secret with the particular recipient. A user, being the only one who keeps his/her private key, can use it to sign a message; the Bitcoin blockchain uses digital signatures for logging transactions on the chain.

In the Bitcoin blockchain, the user's private key is a 256 bit number that is selected at random. The corresponding public key is generated through the use of elliptic curves, in particular the public key is calculated as $K = k * G$, where k is the private key, G is a generator of the cyclic group derived from the elliptic curve and $*$ is the multiplication operation over the `secp256k1` [9] elliptic curve. The generation of the public key from the private is a one-way function, as it is easy to calculate the public key from the private, while it is computationally infeasible to obtain the private key from the public.

Bitcoin addresses that serve as user identifiers are generated from the user public key by successively applying two cryptographic hash functions, namely, `SHA256` and `RIPMED160`, so $A = $ `RIPMED160(SHA256(`K`))`. The generated address is used for applying and logging transactions in the blockchain with the user to whom it belongs.

3.2. *Elliptic curve cryptography*

Cryptographic systems based on elliptic curves (ECC) have been proposed in 1985 by Koblitz [10] and Miller [11] independently as an alternative to conventional public key cryptosystems [12,13].

Their main advantage is that they use smaller parameters compared to the conventional cryptosystems (e.g. RSA). This is due to the apparently increased difficulty of the underlying mathematical problem, the elliptic curve discrete logarithm problem (ECDLP). This problem is believed to require more time for its solution than the time required for the solution of its finite field multiplicative group analog, the discrete logarithm problem (DLP) that ensures the security of a number of cryptosystems (e.g. ElGamal). The security of cryptosystems that rely on discrete logarithms is based on the hypothesis that the DLP is computationally intractable, in the sense that they cannot be computed in polynomial time.

We give the general definition of the elliptic curve. An elliptic curve over the field $\mathbb{F}$ is defined by the Weierstraß equation:

$$E : Y^2 + h(X)Y = f(X),$$

where

$$h(X) = a_1 X + a_3, \quad a_1, a_3 \in \mathbb{F},$$

is a linear polynomial, and

$$f(x) = X^3 + a_2 X^2 + a_4 X + a_6, \quad a_2, a_4, a_6 \in \mathbb{F},$$

is a cubic polynomial, such that there are no solutions $(x, y) \in \overline{\mathbb{F}}^2$ over the algebraic closure $\overline{\mathbb{F}}$ simultaneously satisfying the equations

$$y^2 + h(x)y = f(x), \quad 2y + h(x) = 0 \quad \text{and} \quad h'(x)y = f'(x).$$

In Figures 2 and 3, we plot the elliptic curves derived from the equations $y^2 = x^3 - 3x + 3$ and $y^2 = x^3 - 4x - 1$ over $\mathbb{R}$.

However, in the case of a finite field of prime order, the definition adds up:

An elliptic curve defined over a finite field $\mathbb{F}_p, p > 3$ and prime, is denoted by $E(\mathbb{F}_p)$ and contains the points $(x, y) \in \mathbb{F}_p \times \mathbb{F}_p$ (in affine coordinates) that satisfy the equation (in $\mathbb{F}_p$) $y^2 = x^3 + ax + b$ with $a, b \in \mathbb{F}_p$ satisfying $4a^3 + 27b^2 \neq 0$.

In Figures 4 and 5, we plot the elliptic curves derived from the equations $y^2 = x^3 - 3x + 3$ and $y^2 = x^3 - 4x - 1$ over $\mathbb{F}_p$, $p = 31$.

These points, together with a special point denoted by $\mathcal{O}$, called the point at infinity, and an appropriately defined point addition

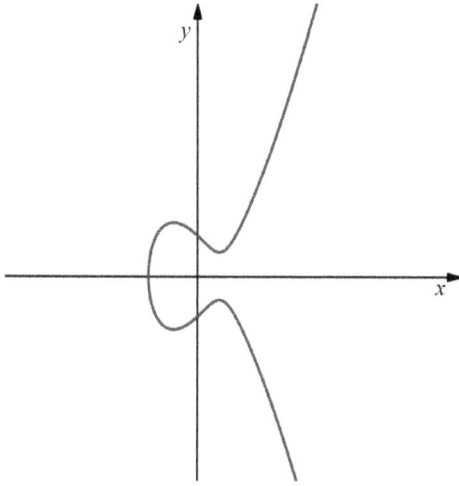

Figure 2. $y^2 = x^3 - 3x + 3$.

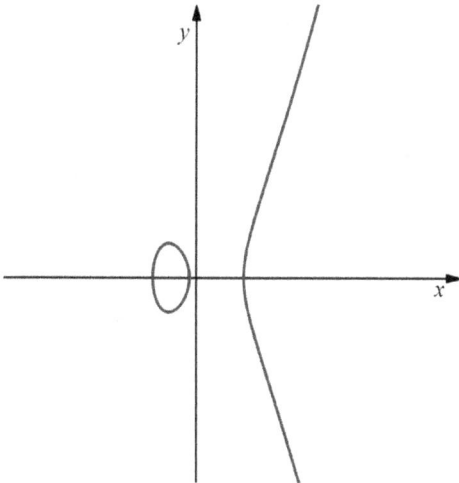

Figure 3. $y^2 = x^3 - 4x - 1$.

operation form an abelian group. This is the elliptic curve group and the point $\mathcal{O}$ is its neutral element.

We are going to give a short description of the group operations on elliptic curves. "Addition" means that given two points and their coordinates, say $P = (x_1, y_1)$ and $Q = (x_2, y_2)$, we have to compute

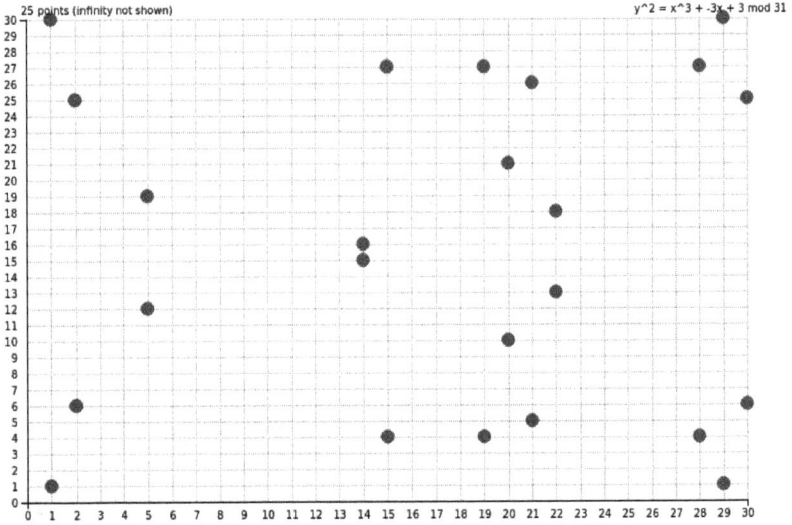

Figure 4. $y^2 \equiv x^3 - 3x + 3 \pmod{31}$.

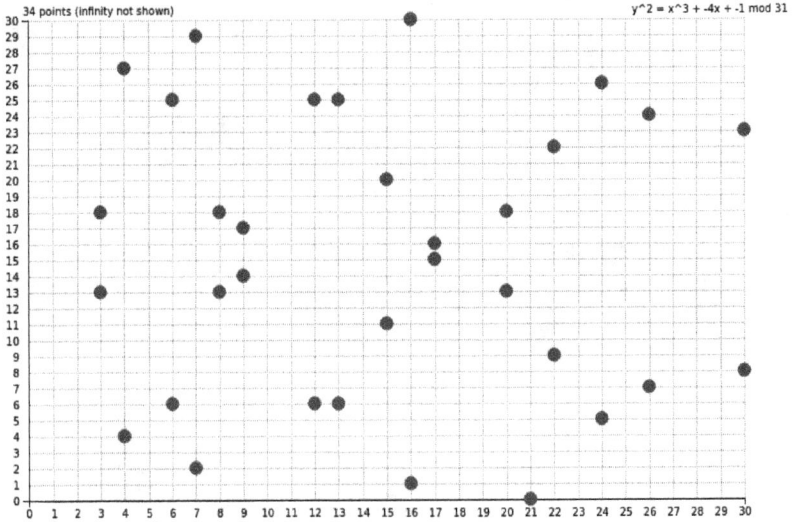

Figure 5. $y^2 \equiv x^3 - 4x - 1 \pmod{31}$.

the coordinates of a third point R such that

$$P + Q = R,$$

$$(x_1, y_1) + (x_2, y_2) = (x_3, y_3).$$

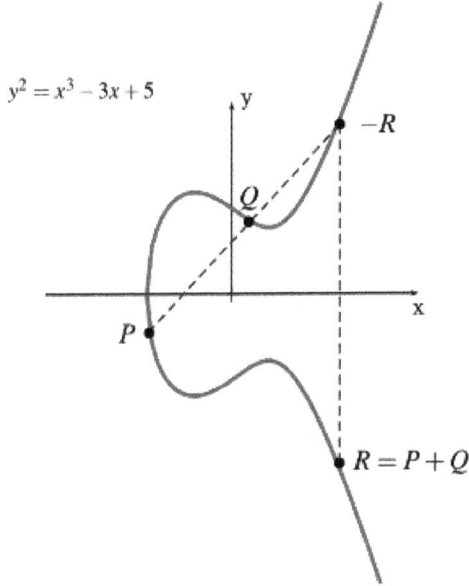

Figure 6. Point addition.

The coordinates of the third point R are given from the following analytical expressions for elliptic curve point addition and point doubling:

$$x_3 \equiv s^2 - x_1 - x_2 \mod p,$$
$$y_3 \equiv s(x_1 - x_3) - y_1 \mod p,$$

where

$$s \equiv \begin{cases} \dfrac{y_2 - y_1}{x_2 - x_1} \mod p & \text{if } P \neq Q \text{ (point addition)}, \\ \dfrac{3x_1^2 + a}{2y_1} \mod p & \text{if } P = Q \text{ (point doubling)}. \end{cases}$$

In Figures 6 and 7, point addition and point doubling are depicted.

The history of elliptic curves dates back to the beginning of the last century. The introduction of the monograph *Guide to Elliptic Curve Cryptography* by Hankerson *et al.* [14] mentioned the following:

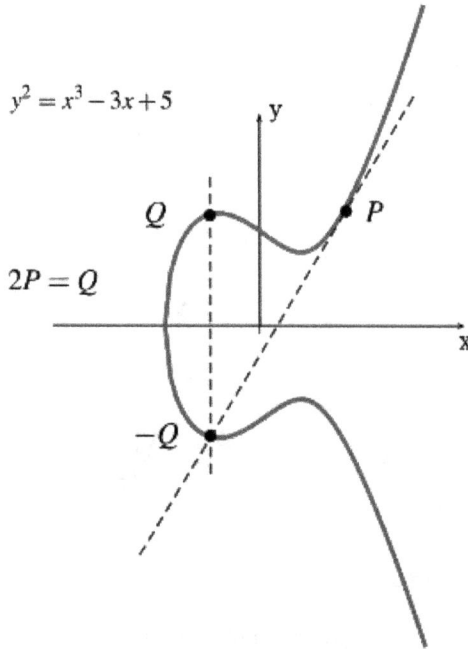

$$y^2 = x^3 - 3x + 5$$

$$2P = Q$$

Figure 7.   Point doubling.

Elliptic curves have a rich and beautiful history, having been studied by mathematicians for over a hundred years. They have been used to solve a diverse range of problems. One example is the congruent number problem that asks for a classification of the positive integers occurring as the area of some right-angled triangle, the lengths of whose sides are rational numbers. Another example is proving Fermat's Last Theorem which states that the equation $x^n + y^n = z^n$ has no nonzero integer solutions for x, y and z when the integer n is greater than 2.

In 1985, Neal Koblitz and Victor Miller independently proposed using elliptic curves to design public-key cryptographic systems. Since then an abundance of research has been published on the security and efficient implementation of elliptic curve cryptography. In the late 1990's, elliptic curve systems started receiving commercial acceptance when accredited standards organizations specified elliptic curve protocols, and private companies included these protocols in their security products.

4. Applications of Blockchains

Distributed ledger technology has already found applications in several domains. Starting with digital cryptocurrencies, where blockchain technology was used to solve the double-spending problem, the technology has since been considered in many other domains, and in some domains, has matured to a point that it is commercially used. The domains where distributed ledgers have been shown to be a good fit include of course financial sector use cases such as the following:

- **Money transfers:** Traditional wire transfers between banks can take days to clear out; equivalent money transfers between customer wallets take minutes or less.
- **Exchanges:** Traditionally, money exchanges incur significant overhead fees; decentralized blockchain-based exchanges that do not require investors to first deposit money with a centralized authority can significantly reduce or entirely diminish relevant fixed costs.
- **Loan processing:** Through the use of smart contracts (concept that was originally popularized with the Ethereum blockchain), it is now possible to significantly reduce delays associated with loan repayments, penalties, etc. by organizing the loan payback processes through the use of such smart contracts that in turn can trigger various actions on behalf of either party.

But of course, blockchain technology has found application (or at least application potential) in a large number of other, diverse use cases, ranging from supply chain management to land registries. In the following few sections, we describe those that have already caught on commercially or that have significant potential.

4.1. *Blockchain for supply chain management and logistics*

One of the main innovations in supply chain management [15] in the last decade, along with the widespread adoption of radio frequency identification (RFID) and Internet of Things (IoT) technologies, is the use of blockchain technologies for managing the flow of information that comes together with the flow of goods and services in logistics. IoT devices and related software expose several vulnerabilities and eventually exploit through the increase of the attack surface of the related information systems that monitor and/or control them. Blockchain technology, being secure by design through the use of the cryptographic protocols it employs, and most of all, through the innovation of the blocks enforcing the security of each other, tackles many of the problems associated with RFID and IoT technologies. The blockchain consensus protocols make it much more difficult for malignant actors to perform successful and harmful cyberattacks. A supply chain monitored with IoT devices and gateways may be used to validate the identities of assets and individuals alike; blockchain technology working on top of that combination can certify who performs what action (for example, which individual signs off the receipt of what goods in a docking station); blockchain acts the security layer that adds trust-worthiness of the entire monitoring/control system. In particular, blockchain can be used to address security challenges related to IP address spoofing and therefore identity management in general, spoofing being the first of the major threat activities recognized by the STRIDE methodology (spoofing, tampering, repudiation, information disclosure, denial of service, elevation of privilege).

4.2.   *Blockchain for games*

A blockchain game is a video game which includes elements that use cryptographic blockchain technologies; a recent survey can be found in Ref. [16]. Blockchain elements in these games are managed by blockchains and can be securely traded (bought/sold/exchanged) between players. The blockchain game therefore is one where in-game transactions are secured by the blockchain mechanisms. For this reason, most major game companies today, including Electronic Arts and Take Two Interactive (owners of Rockstar), have announced plans to integrate blockchain technology in their future games. Gambling, being a particular form of game that involves monetary transactions as the main element of the game, is particularly well suited for blockchain applications. For example, in a virtual casino, if all transactions are recorded on the blockchain, gamblers can immediately see whether the games are fair and the casino pays out in accordance with the stated odds of each game. This is quite different from older protocols devised for the secure exchange of information regarding the outcomes of random experiments: indeed, such protocols are needed to convince players of the fairness of an online game of chance (i.e. that the computer does not cheat when you place a bet on a random event that is however computer-generated). It is a good idea to be mentioned that cryptographic techniques have been used in the past (1980s) for playing games in a long distance relationship [17].

4.3. *Blockchain for land registries and real-estate*

As public records of many countries become digitized, the risks associated with attacks to these infrastructures increase exponentially, with the latest fashion being entire public organizations (hospitals) becoming the victims of ransomware attacks. Blockchain technology, being highly resistant to such attacks (no successful ransomware attack has ever been reported on any blockchain network) offers a way out of this infrastructure security dilemma. Land registries are particularly well suited for the adoption of blockchain technology precisely because the traditional mode of operation has been very bulky and slow, and because chapters (titles, deeds, etc.) can be easily "embedded" in appropriate blockchain-based ledgers. The land registry of the state of Texas, USA, has adopted blockchain technology for its IT infrastructure.

4.4. *Tracking of goods in the food supply chain*

One of the emerging blockchain applications in recent years is food traceability (including food safety and food waste reduction). Relevant issues that are tackled to at least some degree with blockchains include quality assurance, anti-counterfeiting, transparency of origin, fraud prevention, tampering reduction, product recalls, manufacturing collaboration, time reduction from farm to table, and avoidance of intermediaries. Conventional food traceability techniques do not guarantee a satisfactory level of authenticity, reliability, scalability, information integrity and information accuracy. As Yiannas points out [18], the main properties of blockchains, namely, its decentralized, democratic nature based on consensus guaranteeing immutability make the technology ideal for eliminating errors, duplication, redundancy and general inefficiencies in the supply chain as a whole.

5.   The Ethereum Technology

Ethereum [19] is a decentralized public transaction ledger based on blockchain technology. It not only powers the cryptocurrency ETH but also allows the development and deployment of applications that make use of blockchains. This innovative feature of Ethereum opens new possibilities for distributed applications, that are self-governed, whose interactions are governed by the so-called *smart contracts*, which are rules that specify interactions of participating stakeholders. Smart contracts are automatically triggered and executed by the miner nodes and their results are immutably logged into the blockchain. This is the mechanism by which agreement between participating stakeholders is reached.

Ethereum uses the PoW consensus algorithm to validate blocks mined in the network [20], also guarding the blockchain against attacks.

The Ethereum blockchain is guarded against attacks. For a block to get mined, the miner must execute a complex algorithm for solving a difficult problem, as described before. Moreover, the consensus algorithm guards the network from malicious attackers. In order to avoid double spending [21] or attacking the immutability of the blockchain, miners must provide a PoW. Since blocks are chained together, if someone wants to maliciously alter a block, he must redo the calculations and solve the same complex problem for that particular block and all the following (more recent) ones since by changing blocks, the rest of the chain is also affected. Eventually, it becomes clear that it is almost impossible for someone to facilitate that type of malicious action because of the enormous computational power needed in order to perform the same amount of work that went when creating the block and on top of this, it is required to be accepted by everyone in the network. Only by owning the majority of the peers in the network would this kind of attacks be possible, but again the computational power and the costs for block mining make such attacks infeasible.

What makes the Ethereum network so special is that any peer can develop code and deploy it on blockchain, which can then either self-execute or interact with a peer through transactions. The technology that enables transactions and interactions is named smart contract and will be further analyzed in the following sections. Since the Ethereum network enables the use of a Turing-complete language [22], it has raised some criticism with regard to the safety of the entire network and project.

Peers in the network are identified as addresses, which are unique 160-bit identifiers [23]. Peers may exchange ETH or make any type of transactions with the applications deployed on Ethereum. Every transaction performed on the network has a cost, which is measured in gas (unit that measures the amount of computational effort required to execute specific operations on the Ethereum network [24]), and it is paid with ETH from the address that initiates the transaction. The use of gas by the Ethereum network is another measure of protection against malicious users due to the costs it introduces. As a result, attacks such as distributed denial of service (DDoS) are difficult to be launched on the Ethereum network [25], since an attacker has to spend actual money and, in the case of DDoS attacks, has to employ numerous accounts' inevitably, the costs outrun the purpose of the attack.

6.   Prototype Blockchain Application

This section presents a prototype implementation of supply chain workflow management through the use of blockchains (Figure 8). The scenario presented here relates to the transportation of sensitive products that need to meet specific environmental requirements, while clauses specify the financial penalties to be incurred in case of violation of any of the requirements. We assume that during transportation of these products, a number of deployed sensors monitor specific environmental parameters and log the measurements in a blockchain.

Figure 9 depicts the high-level functionalities of the prototype in a top-level use case diagram, which includes four major components: the interface for the stakeholders (manufacturer, retailer,

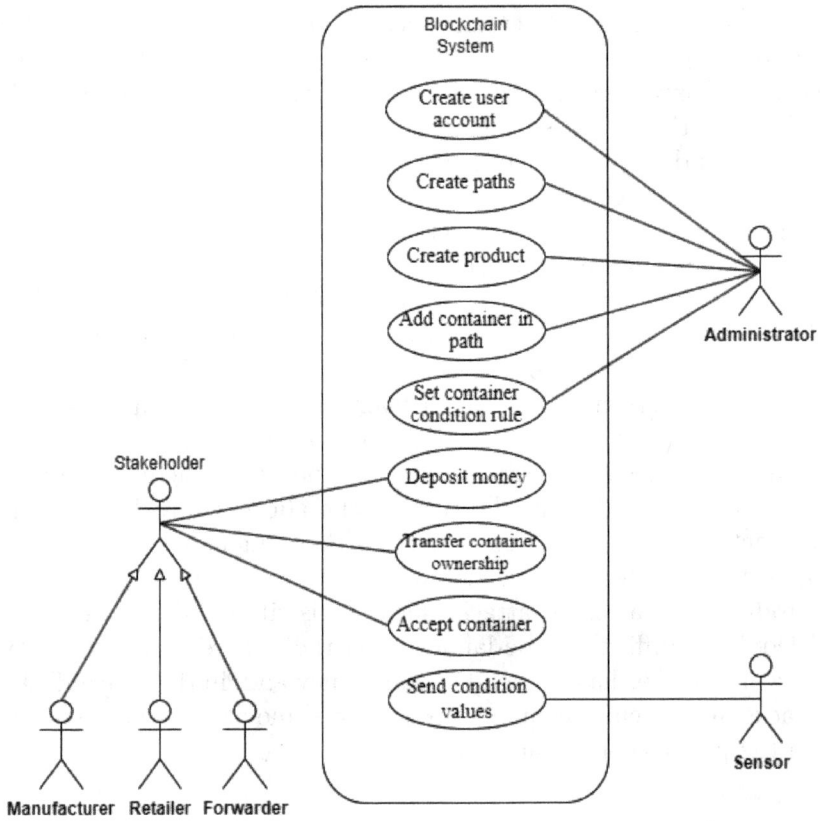

Figure 8. Use case diagram of the application.

forwarder, etc.), the administrator interface, the interface to the Blockchain network, and the interface to the sensors.

After its initial setup by the administrator, the supply chain operates in a self-governed way, that is, at the request of a transport of (sensitive) goods, the stakeholders interact with one another, while sensors log their measurement into the blockchain during the transport and payments are effected automatically with the use of smart contracts that implement the terms of the SLAs between interacting stakeholders. Upon arrival, a smart contract is activated, which checks the agreed transportation requirements and implements the terms of the agreement.

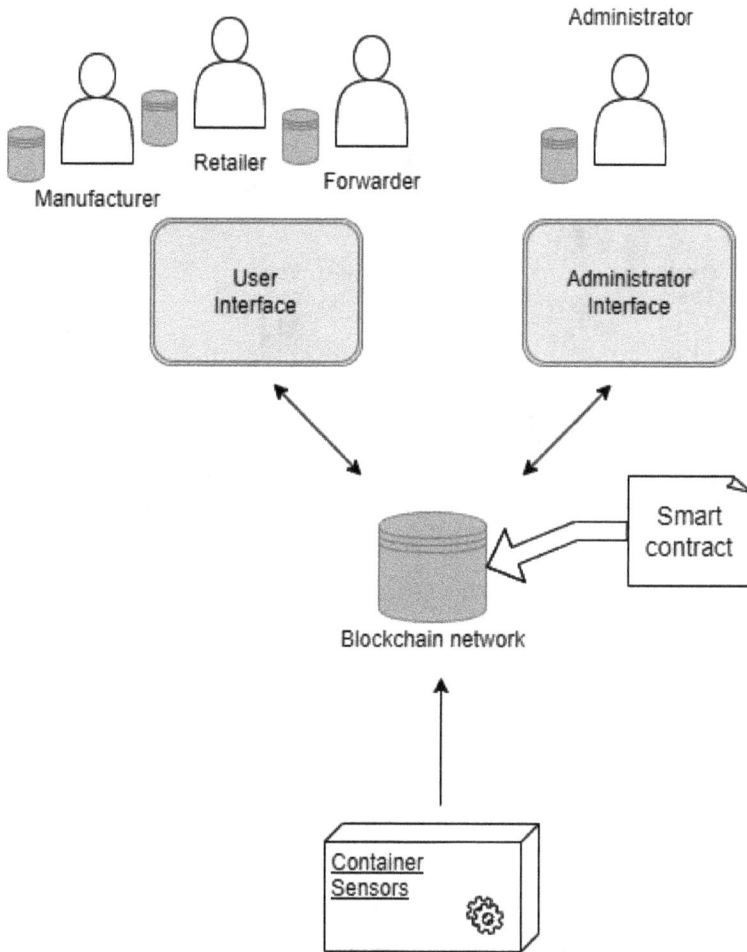

Figure 9.   Use case diagram of the prototype application.

Ethereum-based smart contracts are implemented using the programming language Solidity, which feels familiar to any JEE full-stack developer. All the code developed related to smart contracts in this project was implemented in the aforementioned programming language and at the programming environment of Remix.

Figure 10 shows the architecture of the administrator interface and its main functionalities, which include the ability to create new users, containers and products, and configure parameters for the

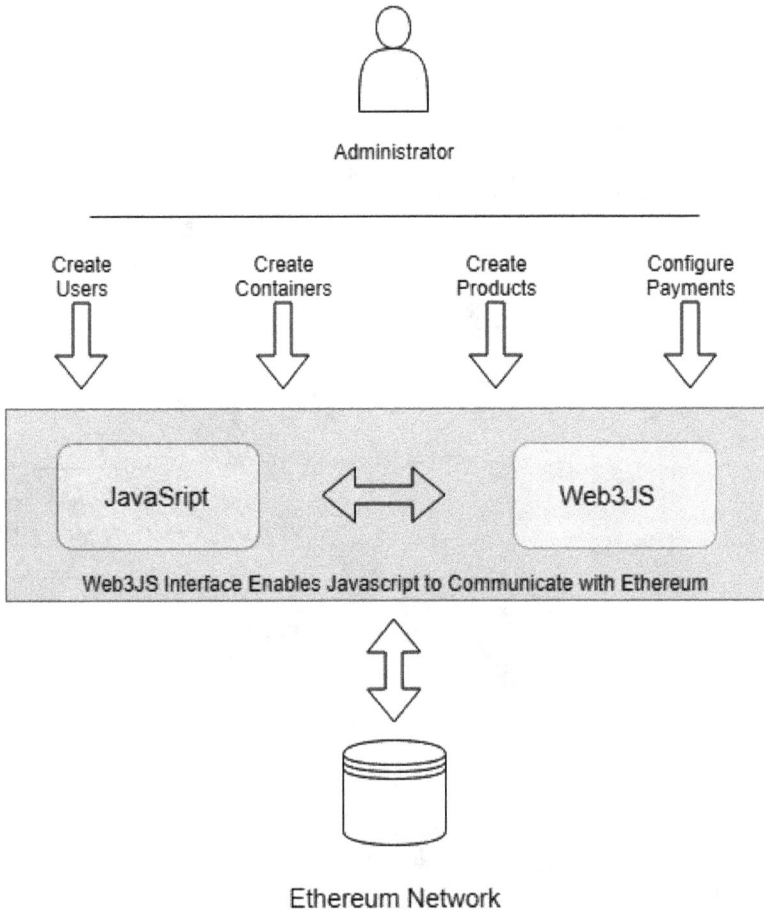

Figure 10. Application administrator functionalities.

application. The admin UI also allows the stakeholder SLAs to be defined.

Figure 11 shows the architecture of the stakeholder interface and its main functionalities. Functionalities include the ability for any of the involved actors to deposit money, transfer ownership (of a product), and finally, to check sensor values.

Figure 12 shows the architecture of the sensor interface, which is responsible for monitoring and recording sensor values in the

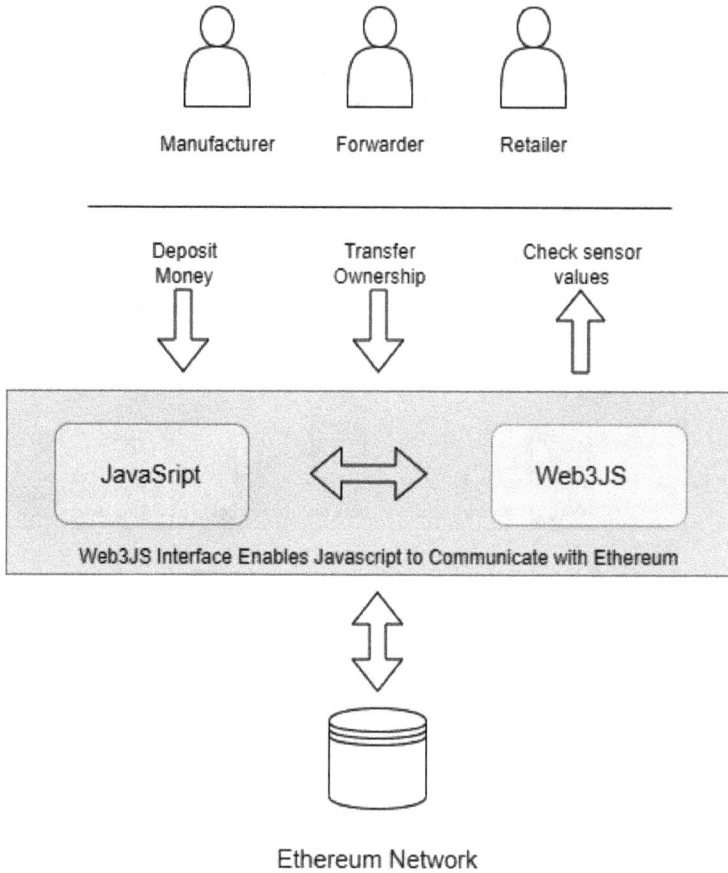

Figure 11. Application stakeholder functionalities.

blockchain: separate threads monitor the readings of the deployed sensors and send the readings (temperature, humidity, acceleration, etc.) of the containers in which they are included to the blockchain.

Figure 13 shows an example usage flow of events in the blockchain. The smart contract code initializes the first block in the blockchain for the particular supply chain. The administrator then creates a new manufacturer account for this blockchain, which is added to the blockchain via the administrator interface. The manufacturer themselves can now deposit an amount for the transfer of the

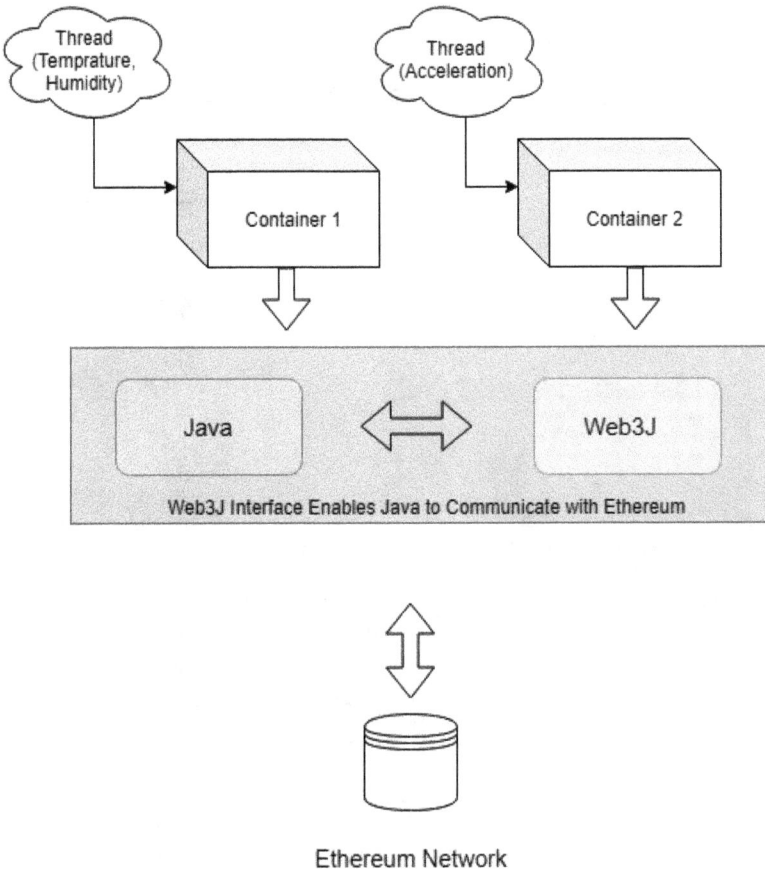

Figure 12.   Sensor interface functionalities.

first container. Afterwards, the sensors in that container continuously read temperature and other data which they submit to the blockchain for permanent and nontampered storage.

All the above functionalities are implemented in a combination of Java and Web3J (https://github.com/web3j/web3j) on the one hand and JavaScript and Web3JS (https://github.com/ChainSafe/web3.js) on the other. It is worth mentioning that the multiple threads needed to continuously read sensor values in an uninterrupted manner are handled by autogenerated functions from the Web3J framework. After reading a sensor value, the responsible thread for the sensor sleeps for a specified interval of time and then

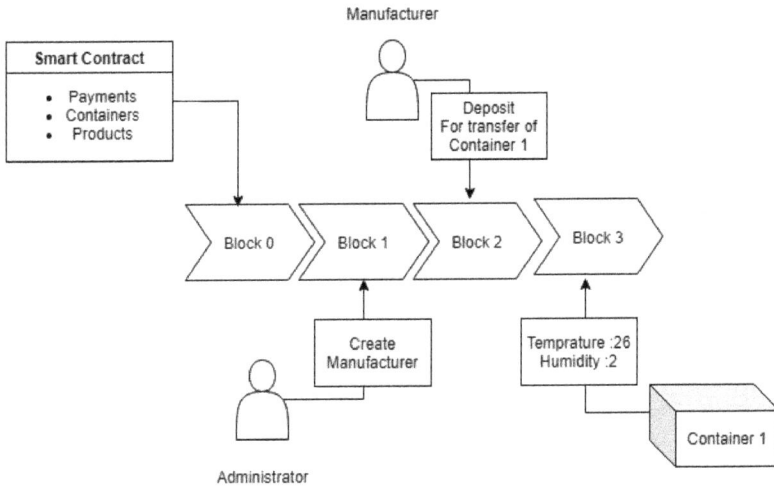

Figure 13. Example flow of events in the application.

reads and sends again the latest value, in an infinite loop, until another thread calls a particular method, "doStop()" method of the thread that sets a volatile flag, which the thread reads on the next iteration of the loop and exits the loop.

6.1. *Escrow accounts*

As an example, the escrow smart contract contains all the information and actions related to payments. In particular, the deposits and the withdrawals are done by this smart contract through its functions. Before initiating the transfer of sensitive products, the final receiver must deposit the money for the payment of each stakeholder. The money, which is in the currency of ETH, is stored in the escrow smart contract. Thus, the smart contract itself acts as the safe third party. The information for payments contains the amount, the amount received, the address of the depositor, the container ID and a Boolean, which represents the completion of payment. The difference between the variables, *amount* and *amount-received*, is that the amount received will be the actual payment to stakeholder, which might get reduced due to the clauses in the SLA. On the other hand, the variable *amount* will be a reference to check how much the initial deposit was. Lastly, the Boolean variable *isPaymentDone* is used in

```
if(_temperatureCheck == 0 && _humcheck == 0 && _accelcheck == 0){
    for(uint i = 0;i< addressToMoney[_receiver].length;i++){
        if(addressToMoney[_receiver][i].containerID == _containerID){
            addressToMoney[_receiver][i].isPaymentDone = true;
        }
    }
    _receiver.transfer(amount);
}
else{
    uint amountPaid = payments[paymentID].amount;
    if(amount-((amount*(_temperatureCheck))/100)-
              ((amount*(_humcheck))/100)-
              ((amount*(_accelcheck))/100)>0){
        amountPaid = amount-((amount*(_temperatureCheck))/100)-
                            ((amount*(_humcheck))/100)-
                            ((amount*(_accelcheck))/100);
        _receiver.transfer(amountPaid);
        payments[paymentID].sender.transfer(
            ((amount*_temperatureCheck)/100)+
            ((amount*_humcheck)/100)+
            ((amount*_accelcheck)/100));
    }
    else{
        amountPaid = 0;
        payments[paymentID].sender.transfer(amount);
    }
}
```

Figure 14.   Smart contract for payments.

order to avoid the double-spending problem. Figure 14 shows the smart contract that implements the service agreement between the consumer and the other stakeholders.

7.   Conclusion

In this chapter, we presented a novel approach for supporting supply chains and interactions among participating stakeholders through blockchain technology. The fundamentals that underlie blockchains are covered along with available technologies. A prototype application is presented as a proof of concept of the proposed approach to show that blockchains are a natural and intuitive option for supporting supply chain business processes and service agreements among stakeholders in the framework of the Ethereum network. The prototype runs every Use-Case very fast in the order of seconds. This was expected, as the algorithms implementing each of the Use-Cases described in this chapter have theoretical constant time complexity.

The fact that the response times are not smaller is due to the network overheads that any distributed system such as blockchains suffer, especially when the application services are not directly connected to high-speed (backbone Internet) hubs.

As future research, we plan to investigate the possibility to replace some or even all of the current consensus mechanisms that are based on PoW, which are criticized for not being environmentally friendly, with other secure cryptographic mechanisms that can still guarantee all the major security and transparency aspects of blockchains.

References

[1] I. Kozlenkova, G. T. M. Hult, D. Lund, J. Mena, and P. Kekec (2015). The role of marketing channels in supply chain management. *J. Retail.*, **91**, 586–609. doi: 10.1016/j.jretai.2015.03.003.

[2] www.cyrene.eu.

[3] https://eur-lex.europa.eu/eli/reg/2019/881/oj.

[4] https://www.enisa.europa.eu/publications/cybersecurity-certification-eucc-candidate-scheme.

[5] L. Lamport, R. Shostak, and M. Pease (1982). The byzantine generals problem, *ACM Trans. Program. Lang. Sys.*, **4**, 382–401.

[6] A. Narayanan, J. Bonneau, E. Felten, A. Miller, and S. Goldfeder (2016). *Bitcoin and Cryptocurrency Technologies* (Princeton University Press).

[7] S. Nakamoto (2009). Bitcoin: A peer-to-peer electronic cash system. http://www.bitcoin.org/bitcoin.pdf.

[8] W. Diffie and M. E. Hellman (1976). New directions in cryptography. *IEEE Trans. Inform. Theory*, **22**(6), 644–654.

[9] Standards for efficient cryptography, Certicom research. https://www.secg.org/sec2-v2.pdf.

[10] N. Koblitz (1987). Elliptic curve cryptosystems. *Math. Comp.*, **48**(177), 203–209.

[11] V. S. Miller (1986). Use of elliptic curves in cryptography. In *Advances in Cryptology — CRYPTO '85 Proceedings*, (Springer, Berlin, Heidelberg), pp. 417–426.

[12] J. H. Silverman (1986). *The Arithmetic of Elliptic Curves* (Springer-Verlag, New York Inc).

[13] L. Washington (2003). *Elliptic Curves: Number Theory and Cryptography* (Chapman & Hall/CRC, Boca Raton, London, New York, Washington D.C.).

[14] D. R. Hankerson, A. Menezes, and S. Vanstone (2004). *Guide to Elliptic Curve Cryptography* (Springer-Verlag, New York Inc).

[15] I. T. Christou (2012). *Quantitative Methods in Supply Chain Management: Models and Algorithms* (Springer-Verlag, London Limited).

[16] T. Min, H. Wang, Y. Guo, and W. Cai (2019). Blockchain games: A survey. In *Proceedings of the IEEE Conference on Games (CoG)*, pp. 1–8, (20–23 August 2019, London, UK).

[17] M. Blum (1981). Coin flipping by telephone: A protocol for solving impossible problems. In *Advances in Cryptology, CRYPTO 81*, pp. 11–15.

[18] F. Yinnas (2018). A new era of food transparency powered by blockchain. *Innovations*, **12**, 46–56.

[19] https://ethereum.org.

[20] https://ethereum.org/en/developers/docs/consensus-mechanisms/pow.

[21] H. T. M. Gamage, H. Weerasinghe, and N. G. J. Dias (2020). A survey on blockchain technology concepts, applications, and issues, *SN Comput. Sci.*, **1**, 114.

[22] M. R. Garey and D. S. Johnson (1979). *Computers and Intractability: A Guide to the Theory of NP-Completeness* (first edition). In *Series of Books in the Mathematical Sciences*, W. H. Freeman and Company, New York, http://www.amazon.com/Computers-Intractability-NP-Completeness-Mathematical-Sciences/dp/0716710455.

[23] A. Antonopoulos (2017). *Mastering Bitcoin: Programming the Open Blockchain* (O'Reilly Media, Sebastopol, California, https://www.oreilly.com). https://books.google.gr/books?id=MpwnDwAAQBAJ.

[24] https://ethereum.org/en/developers/docs/gas.

[25] U. Javaid, A. Siang, M. Aman, and B. Sikdar (2018). Mitigating IoT device based DDoS attacks using blockchain. In *CryBlock '18: Proceedings of the 1st Workshop on Cryptocurrencies and Blockchains for Distributed Systems*, pp. 71–76. doi: 10.1145/3211933.3211946.

© 2023 World Scientific Publishing Company
https://doi.org/10.1142/9789811271922_0003

Chapter 3

Consecutive Square-Free Numbers $[n^c \tan^\theta (\log n)]$, $[n^c \tan^\theta (\log n)] + 1$

Stoyan Ivanov Dimitrov

Faculty of Applied Mathematics and Informatics,
Technical University of Sofia,
Blvd. St. Kliment Ohridski 8,
Sofia 1756, Bulgaria

sdimitrov@tu-sofia.bg

Let $[\,\cdot\,]$ be the floor function. In this chapter, we prove that when $1 < c < \frac{7}{6}$ and $\theta > 1$ is fixed, then there exist infinitely many square-free pairs of the form $[n^c \tan^\theta (\log n)]$, $[n^c \tan^\theta (\log n)] + 1$.

1. Notations

Let x be a sufficiently large positive number. By ε, we denote an arbitrary small positive constant, not the same in all appearances. We denote by $\mu(n)$ the Möbius function and by $\tau(n)$ the number of positive divisors of n. As usual, $[t]$ and $\{t\}$ denote the integer part, respectively, the fractional part of t. Moreover, $e(y) = e^{2\pi i y}$. Instead of $m \equiv n \pmod{k}$, we write for simplicity $m \equiv n(k)$. Throughout this chapter, we suppose that $1 < c < \frac{7}{6}$ and $\gamma = \frac{1}{c}$. Assume that $\theta > 1$ is fixed. Denote

$$z = x^{\frac{2c-1}{4}}, \tag{1}$$

$$\psi(t) = \{t\} - 1/2, \tag{2}$$

$$\Delta_1 = e^{\pi \left[\frac{\log x}{\pi}\right] + \arctan 1}, \tag{3}$$

$$\Delta_2 = e^{\pi \left[\frac{\log x}{\pi}\right] + \arctan 2}, \tag{4}$$

$$S_c(x) = \sum_{\Delta_1 \leq n < \Delta_2} \mu^2\big([n^c \tan^\theta (\log n)]\big) \mu^2\big([n^c \tan^\theta (\log n)] + 1\big). \tag{5}$$

2. Introduction and Statement of the Result

In 1932, Carlitz [3] showed that there exist infinitely many pairs of consecutive square-free numbers. More precisely, he proved the asymptotic formula

$$\sum_{n \leq x} \mu^2(n)\mu^2(n+1) = \prod_p \left(1 - \frac{2}{p^2}\right) x + \mathcal{O}(x^{\lambda+\varepsilon}), \tag{6}$$

where $\lambda = 2/3$. Afterwards, the reminder term of (6) was improved by Mirsky [7] and Heath-Brown [6]. The best result up to now belongs to Reuss [8] with $\lambda = (26 + \sqrt{433})/81$. Another interesting problem we know in number theory is square-free numbers of the form $[n^c]$. In 1975, Stux [10] was the first to study the distribution of such numbers. He proved that for any fixed $1 < c < \frac{4}{3}$, the asymptotic formula

$$\sum_{n \leq x} \mu^2([n^c]) = \frac{6}{\pi^2}x + \Delta \tag{7}$$

holds. Here,

$$\Delta = o(x).$$

Subsequently, the result of Stux was improved by Rieger [9] and Cao and Zhai [1]. In 2008, Cao and Zhai [2] improved their result by showing that for any fixed $1 < c < \frac{149}{87}$, $\gamma = c^{-1}$ and $0 < \varepsilon_0 < (149\gamma - 87)/400$, the asymptotic formula (7) holds with

$$\Delta = \mathcal{O}\left(x^{1-\varepsilon_0}\right),$$

and this is the best result up to now.

Recently, the author [4] proved that there exist infinitely many prime numbers of the form $[n^c \tan^\theta (\log n)]$. More precisely, we showed that when $1 < c < \frac{12}{11}$ and $\theta > 1$ is fixed, then the lower bound

$$\sum_{\substack{\Delta_1 \le n < \Delta_2 \\ [n^c \tan^\theta (\log n)] = p}} 1 \gg \frac{x}{\log x}$$

holds. Here, Δ_1 and Δ_2 are denoted by (3) and (4).

Motivated by these results, we prove that there exist infinitely many square-free pairs of the form $[n^c \tan^\theta (\log n)]$, $[n^c \tan^\theta (\log n)]+1$. More precisely, we establish the following theorem.

Theorem 1. *Let $1 < c < \frac{7}{6}$ and $\theta > 1$ be a fixed. Then for the sum (5), the lower bound*

$$S_c(x) \gg x \tag{8}$$

holds.

3. Preliminary Lemmas

Lemma 1. *Let $H \ge 1$. Then for the function defined by (2), we have*

$$\psi(t) = \sum_{1 \le |h| \le H} a(h) e(ht) + \mathcal{O}\left(\sum_{|h| \le H} b(h) e(ht) \right),$$

where

$$a(h) \ll \frac{1}{|h|}, \quad b(h) \ll \frac{1}{H}. \tag{9}$$

Proof.   See Ref. [11].   □

Lemma 2. *Let $q \geq 0$ be an integer. Suppose that $f(t)$ has $q + 2$ continuous derivatives on I and that $I \subseteq (N, 2N]$. Assume also that there is some constant F such that*

$$|f^{(r)}(t)| \asymp FN^{-r} \tag{10}$$

for $r = 1, \ldots, q + 2$. Let $Q = 2^q$. Then

$$\left| \sum_{n \in I} e(f(n)) \right| \ll F^{\frac{1}{4Q-2}} N^{1-\frac{q+2}{4Q-2}} + F^{-1}N.$$

The implied constant depends only upon the implied constants in (10).

Proof. See Ref. ([5], Theorem 2.9). □

4. Beginning of the Proof

Using (1), (5) and the well-known identity $\mu^2(n) = \sum_{d^2 | n} \mu(d)$, we write

$$S_c(x) = \sum_{\substack{\Delta_1 \leq n < \Delta_2 \\ d^2 | [n^c \tan^\theta (\log n)]}} \sum \mu(d) \sum_{t^2 | [n^c \tan^\theta (\log n)] + 1} \mu(t)$$

$$= \sum_{\substack{d,t \\ (d,t)=1}} \mu(d)\mu(t) \sum_{\substack{\Delta_1 \leq n < \Delta_2 \\ [n^c \tan^\theta (\log n)] \equiv 0 \, (d^2) \\ [n^c \tan^\theta (\log n)] + 1 \equiv 0 \, (t^2)}} 1$$

$$= S_c^{(1)}(x) + S_c^{(2)}(x), \tag{11}$$

where

$$S_c^{(1)}(x) = \sum_{\substack{dt \leq z \\ (d,t)=1}} \mu(d)\mu(t) \sum_{\substack{\Delta_1 \leq n < \Delta_2 \\ [n^c \tan^\theta (\log n)] \equiv 0 \, (d^2) \\ [n^c \tan^\theta (\log n)] + 1 \equiv 0 \, (t^2)}} 1, \tag{12}$$

$$S_c^{(2)}(x) = \sum_{\substack{dt > z \\ (d,t)=1}} \mu(d)\mu(t) \sum_{\substack{\Delta_1 \leq n < \Delta_2 \\ [n^c \tan^\theta (\log n)] \equiv 0 \, (d^2) \\ [n^c \tan^\theta (\log n)] + 1 \equiv 0 \, (t^2)}} 1. \tag{13}$$

5. Estimation of $S_c^{(1)}(x)$

From (12), we obtain

$$
S_c^{(1)}(x) = \sum_{\substack{dt \le z \\ (d,t)=1}} \mu(d)\mu(t) \sum_{\substack{(\Delta_1^c \tan^\theta (\log \Delta_1)-1)d^{-2} < k < \Delta_2^c \tan^\theta (\log \Delta_2)d^{-2} \\ kd^2+1\equiv 0 \,(t^2)}}
$$

$$
\times \sum_{\substack{\Delta_1 \le n < \Delta_2 \\ [n^c \tan^\theta (\log n)]=kd^2}} 1
$$

$$
= \sum_{\substack{dt \le z \\ (d,t)=1}} \mu(d)\mu(t) \sum_{\substack{(\Delta_1^c \tan^\theta (\log \Delta_1)-1)d^{-2} < k < \Delta_2^c \tan^\theta (\log \Delta_2)d^{-2} \\ kd^2+1\equiv 0 \,(t^2)}}
$$

$$
\times \sum_{\substack{\Delta_1 \le n < \Delta_2 \\ kd^2 \le n^c \tan^\theta (\log n) < kd^2+1}} 1
$$

$$
= \sum_{\substack{dt \le z \\ (d,t)=1}} \mu(d)\mu(t) \sum_{\substack{(\Delta_1^c \tan^\theta (\log \Delta_1)-1)d^{-2} < k < \Delta_2^c \tan^\theta (\log \Delta_2)d^{-2} \\ kd^2+1\equiv 0 \,(t^2)}}
$$

$$
\times \sum_{\Delta_1' \le n < \Delta_2'} 1 + \mathcal{O}(x^\varepsilon)
$$

$$
= \sum_{\substack{dt \le z \\ (d,t)=1}} \mu(d)\mu(t) \sum_{\substack{(\Delta_1^c \tan^\theta (\log \Delta_1)-1)d^{-2} < k < \Delta_2^c \tan^\theta (\log \Delta_2)d^{-2} \\ kd^2+1\equiv 0 \,(t^2)}}
$$

$$
\times \left([-\Delta_k'] - [-\Delta_k''] \right) + \mathcal{O}(x^\varepsilon)
$$

$$
= S_c^{(3)}(x) + S_c^{(4)}(x) + \mathcal{O}(x^\varepsilon), \tag{14}
$$

where

$$
[\Delta_k', \Delta_k'') \subset [\Delta_1, \Delta_2), \tag{15}
$$

and the interval $[\Delta_k', \Delta_k'')$ is a solution of the system inequalities

$$
\left| \begin{array}{l} n^c \tan^\theta (\log n) \ge kd^2, \\ n^c \tan^\theta (\log n) < kd^2 + 1, \end{array} \right. \tag{16}
$$

and where

$$S_c^{(3)}(x) = \sum_{\substack{dt \leq z \\ (d,t)=1}} \mu(d)\mu(t) \sum_{\substack{(\Delta_1^c \tan^\theta(\log \Delta_1)-1)d-2 < k < \Delta_2^c \tan^\theta(\log \Delta_2)d-2 \\ kd^2+1 \equiv 0 \, (t^2)}}$$

$$\times \left(\Delta_k'' - \Delta_k'\right), \tag{17}$$

$$S_c^{(4)}(x) = \sum_{\substack{dt \leq z \\ (d,t)=1}} \mu(d)\mu(t) \sum_{\substack{(\Delta_1^c \tan^\theta(\log \Delta_1)-1)d-2 < k < \Delta_2^c \tan^\theta(\log \Delta_2)d-2 \\ kd^2+1 \equiv 0 \, (t^2)}}$$

$$\times \left(\psi\left(-\Delta_k''\right) - \psi\left(-\Delta_k'\right)\right). \tag{18}$$

5.1. *Lower bound for $S_c^{(3)}(x)$*

Consider the function $t(y)$ defined by

$$t = y^c \tan^\theta(\log y) \quad \text{for } y \in \left[\Delta_1', \Delta_2'\right]. \tag{19}$$

Using (16), (17) and (19), Abel's summation formula and the mean-value theorem, we get

$$S_c^{(3)}(x)$$

$$= \sum_{\substack{dt \leq z \\ (d,t)=1}} \mu(d)\mu(t) \sum_{\substack{(\Delta_1^c \tan^\theta(\log \Delta_1)-1)d-2 < k < \Delta_2^c \tan^\theta(\log \Delta_2)d-2 \\ kd^2+1 \equiv 0 \, (t^2)}}$$

$$\times \left(y(kd^2 + 1) - y(kd^2)\right)$$

$$= \sum_{\substack{dt \leq z \\ (d,t)=1}} \mu(d)\mu(t) \left[\left(y(\Delta_2^c \tan^\theta(\log \Delta_2) + 1) - y(\Delta_2^c \tan^\theta(\log \Delta_2))\right)\right.$$

$$\times \sum_{\substack{(\Delta_1^c \tan^\theta(\log \Delta_1)-1)d-2 < k < \Delta_2^c \tan^\theta(\log \Delta_2)d-2 \\ kd^2+1 \equiv 0 \, (t^2)}} 1$$

$$- \int_{(\Delta_1^c \tan^\theta(\log \Delta_1)-1)d-2}^{\Delta_2^c \tan^\theta(\log \Delta_2)d-2} \left(\sum_{\substack{(\Delta_1^c \tan^\theta(\log \Delta_1)-1)d-2 < k \leq u \\ kd^2+1 \equiv 0 \, (t^2)}} 1\right)$$

$$\times \frac{\partial\left(y(ud^2 + 1) - y(ud^2)\right)}{\partial u} \, du\Bigg]$$

$$= \sum_{\substack{dt \le z \\ (d,t)=1}} \mu(d)\mu(t)$$

$$\times \left[y'(\xi_1) \left(\frac{\Delta_2^c \tan^\theta (\log \Delta_2) - \Delta_1^c \tan^\theta (\log \Delta_1) + 1}{d^2 t^2} + \mathcal{O}(1) \right) \right.$$

$$- \int_{(\Delta_1^c \tan^\theta (\log \Delta_1)-1)d^{-2}}^{\Delta_2^c \tan^\theta (\log \Delta_2)d^{-2}} \left(\frac{u - (\Delta_1^c \tan^\theta (\log \Delta_1) - 1)d^{-2}}{t^2} + \mathcal{O}(1) \right)$$

$$\left. \times \frac{\partial \big(y(ud^2 + 1) - y(ud^2) \big)}{\partial u} \, du \right]$$

$$= \sum_{\substack{dt \le z \\ (d,t)=1}} \frac{\mu(d)\mu(t)}{t^2} \int_{(\Delta_1^c \tan^\theta (\log \Delta_1)-1)d^{-2}}^{\Delta_2^c \tan^\theta (\log \Delta_2)d^{-2}} \big(y(ud^2 + 1) - y(ud^2) \big) \, du$$

$$+ \mathcal{O}\left(z x^\varepsilon \big(|y'(\xi_1)| + |y'(\xi_2)| \big) \right)$$

$$= \sum_{\substack{dt \le z \\ (d,t)=1}} \frac{\mu(d)\mu(t)}{d^2 t^2} \int_{\Delta_1^c \tan^\theta (\log \Delta_1)-1}^{\Delta_2^c \tan^\theta (\log \Delta_2)} \big(y(z+1) - y(z) \big) \, dz$$

$$+ \mathcal{O}\left(z x^\varepsilon \big(|y'(\xi_1)| + |y'(\xi_2)| \big) \right)$$

$$= \sum_{\substack{dt \le z \\ (d,t)=1}} \frac{\mu(d)\mu(t)}{d^2 t^2} \int_{\Delta_1^c \tan^\theta (\log \Delta_1)-1}^{\Delta_2^c \tan^\theta (\log \Delta_2)} y'(\zeta_z) \, dz$$

$$+ \mathcal{O}\left(z x^\varepsilon \big(|y'(\xi_1)| + |y'(\xi_2)| \big) \right), \tag{20}$$

where

$$\begin{vmatrix} y(\xi_1) \asymp \xi_1^\gamma \\ y(\xi_2) \asymp \xi_2^\gamma \\ y(\zeta_z) \asymp \zeta_z^\gamma \\ z < \zeta_z < z + 1 \\ \Delta_2^c \tan^\theta (\log \Delta_2) < \xi_1 < \Delta_2^c \tan^\theta (\log \Delta_2) + 1 \\ \Delta_1^c \tan^\theta (\log \Delta_1) - 1 < \xi_2 < \Delta_1^c \tan^\theta (\log \Delta_1). \end{vmatrix} \tag{21}$$

For the implicit function y defined by (19), the first derivative with respect to t is

$$y' = \frac{y^{1-c}}{\left(c\tan(\log y) + \theta\sec^2(\log y)\right)\tan^{\theta-1}(\log y)}.\qquad(22)$$

Now, (1), (3), (4), (15), (19)–(21) and the well-known asymptotic formula

$$\sum_{\substack{dt\leq z\\(d,t)=1}} \frac{\mu(d)\mu(t)}{d^2t^2} = \prod_p\left(1-\frac{2}{p^2}\right) + \mathcal{O}\left(z^{\varepsilon-1}\right)$$

gives us

$$S_c^{(3)}(x) \gg x.\qquad(23)$$

5.2.   Upper bound for $S_c^{(4)}(x)$

By (18), we have

$$S_c^{(4)}(x) \ll \sum_{dt\leq z}\left|S_c^{(5)}(x)\right|,\qquad(24)$$

where

$$S_c^{(5)}(x) = \sum_{\substack{(\Delta_1^c\tan^\theta(\log\Delta_1)-1)d^{-2}<k<\Delta_2^c\tan^\theta(\log\Delta_2)d^{-2}\\kd^2+1\equiv0\,(t^2)}}\left(\psi\left(-\Delta_k''\right)-\psi\left(-\Delta_k'\right)\right).\qquad(25)$$

Now, (2), (25) and Lemma 1 yield

$$S_c^{(5)}(x) = S_c^{(6)}(x) + S_c^{(7)}(x) + S_c^{(8)}(x),\qquad(26)$$

where

$$S_c^{(6)}(x) = \sum_{\substack{(\Delta_1^c\tan^\theta(\log\Delta_1)-1)d^{-2}<k<\Delta_2^c\tan^\theta(\log\Delta_2)d^{-2}\\kd^2+1\equiv0\,(t^2)}}\sum_{1\leq|h|\leq H}a(h)$$

$$\times\left(e\left(-h\Delta_k''\right)-e\left(-h\Delta_k'\right)\right),\qquad(27)$$

$$S_c^{(7)}(x) \ll \sum_{\substack{(\Delta_1^c \tan^\theta(\log \Delta_1)-1)d-2 < k < \Delta_2^c \tan^\theta(\log \Delta_2)d-2 \\ kd^2+1 \equiv 0 \, (t^2)}}$$

$$\times \sum_{|h| \leq H} b(h) e\left(-h\Delta_k'\right), \tag{28}$$

$$S_c^{(8)}(x) \ll \sum_{\substack{(\Delta_1^c \tan^\theta(\log \Delta_1)-1)d-2 < k < \Delta_2^c \tan^\theta(\log \Delta_2)d-2 \\ kd^2+1 \equiv 0 \, (t^2)}}$$

$$\times \sum_{|h| \leq H} b(h) e\left(-h\Delta_k''\right). \tag{29}$$

Put

$$H = \frac{x^{\frac{2c-1}{4}}}{dt}. \tag{30}$$

We first estimate $S_c^{(7)}(x)$. From (3), (4), (9) and (28), we deduce

$$S_c^{(7)}(x)$$

$$\ll \frac{1}{H} \sum_{|h| \leq H} \left| \sum_{\substack{(\Delta_1^c \tan^\theta(\log \Delta_1)-1)d-2 < k < \Delta_2^c \tan^\theta(\log \Delta_2)d-2 \\ kd^2+1 \equiv 0 \, (t^2)}} e\left(-h\Delta_k'\right) \right|$$

$$\ll \frac{x^c}{Hd^2t^2} + \frac{1}{H} \sum_{1 \leq h \leq H} \left| \sum_{\substack{(\Delta_1^c \tan^\theta(\log \Delta_1)-1)d-2 < k < \Delta_2^c \tan^\theta(\log \Delta_2)d-2 \\ kd^2+1 \equiv 0 \, (t^2)}} \right.$$

$$\left. \times \, e\left(-h\Delta_k'\right) \right|$$

$$\ll \frac{x^c}{Hd^2t^2} + \frac{1}{H} \sum_{1 \leq h \leq H} \left| \sum_{\Delta_1^c \tan^\theta(\log \Delta_1)t-2 < l \leq (\Delta_2^c \tan^\theta(\log \Delta_2)+1)t-2} \right.$$

$$\left. \times \, e\left(-h\Delta_{(lt^2-1)d-2}'\right) \right|. \tag{31}$$

It is easy to see that (3) and (4) imply

$$\frac{2\left(\Delta_1^c \tan^\theta(\log \Delta_1)\right)}{t^2} < \frac{\Delta_2^c \tan^\theta(\log \Delta_2) + 1}{t^2}. \tag{32}$$

Taking into account (31) and (32), we split the range of l into dyadic subintervals to obtain

$$S_c^{(7)}(x) \ll \frac{x^c}{Hd^2t^2} + \frac{\log x}{H} \sum_{1 \le h \le H} \left| \sum_{L < l \le 2L} e\left(-h\Delta'_{(lt^2-1)d^{-2}}\right) \right|, \tag{33}$$

where

$$\begin{vmatrix} L \asymp \frac{x^c}{t^2} \\ \frac{\Delta_1^c \tan^\theta(\log \Delta_1)}{t^2} \le L \le \frac{\Delta_2^c \tan^\theta(\log \Delta_2) + 1}{2t^2}. \end{vmatrix} \tag{34}$$

According to (15), (16), (33) and (34), the variable $\Delta'_{(lt^2-1)d^{-2}}$ is an implicit function of l defined by

$$y^c \tan^\theta(\log y) = lt^2 - 1 \quad \text{for } l \in (L, 2L] \tag{35}$$

and

$$y \subset [\Delta_1, \Delta_2). \tag{36}$$

Bearing in mind (3), (4), (19), (22), (34)–(36), we find

$$y' \asymp L^{\gamma-1}t^{2\gamma}. \tag{37}$$

Proceeding in the same way, we derive

$$y'' \asymp L^{\gamma-2}t^{2\gamma}. \tag{38}$$

Now (33), (34), (37), (38) and Lemma 2 with $q = 0$ lead to

$$S_c^{(7)}(x) \ll \frac{x^c}{Hd^2t^2} + \frac{\log x}{H} \sum_{1 \le h \le H} \left(h^{\frac{1}{2}}L^{\frac{\gamma}{2}}t^\gamma + h^{-1}L^{1-\gamma}t^{-2\gamma}\right)$$

$$\ll x^c H^{-1}d^{-2}t^{-2} + L^{\frac{\gamma}{2}}H^{\frac{1}{2}}t^\gamma \log x + L^{1-\gamma}H^{-1}t^{-2\gamma}\log^2 x$$

$$\ll x^c H^{-1}d^{-2}t^{-2} + x^{\frac{1}{2}}H^{\frac{1}{2}} + x^{c-1}H^{-1}t^{-2}\log^2 x$$

$$\ll x^c H^{-1}d^{-2}t^{-2} + x^{\frac{1}{2}}H^{\frac{1}{2}}. \tag{39}$$

Arguing in the same way for the sum defined by (29), we get

$$S_c^{(8)}(x) \ll x^c H^{-1} d^{-2} t^{-2} + x^{\frac{1}{2}} H^{\frac{1}{2}}. \tag{40}$$

The sum $S_c^{(6)}(K)$ remains to be estimated. Working similar to the estimation of $S_c^{(7)}(x)$ from (9) and (27), we have

$$S_c^{(6)}(x) \ll \sum_{1 \le |h| \le H} \frac{1}{h} \left| \sum_{L < l \le 2L} \Phi_h(l) e\left(- h \Delta'_{(lt^2-1)d^{-2}} \right) \right|, \tag{41}$$

where L satisfies (34) and

$$\Phi_h(u) = e\left(h(\Delta'_{(ut^2-1)d^{-2}} - \Delta''_{(ut^2-1)d^{-2}}) \right) - 1. \tag{42}$$

From (3), (4), (19), (22), (34), (35), (36), (42) and the mean-value theorem, that it follows

$$|\Phi_h(u)| \le 2 \left| \sin\left(\pi h(\Delta'_u - \Delta''_u)\right) \right| \ll |h| |\Delta'_u - \Delta''_u| \ll |h| x^{1-c}, \tag{43}$$

$$\Phi'_h(u) = 2\pi i h \left(\frac{\partial \Delta'_u}{\partial u} - \frac{\partial \Delta''_u}{\partial u} \right) e\left(h(\Delta'_u - \Delta''_u) \right) \ll |h| x^{1-2c} \tag{44}$$

for $u \in [L, 2L]$. Now, (34), (35), (41), (43), (44) and Abel's summation formula give us

$$S_c^{(6)}(x) \ll \sum_{1 \le |h| \le H} \frac{1}{h} \left| \Phi_h(2L) \sum_{L < l \le 2L} e\left(- h \Delta'_{(lt^2-1)d^{-2}} \right) \right|$$

$$+ \sum_{1 \le |h| \le H} \frac{1}{h} \int_L^{2L} \left| \Phi'_h(u) t^2 \sum_{L < l \le u} e\left(- h \Delta'_{(lt^2-1)d^{-2}} \right) \right| du$$

$$\ll x^{1-c} |S_c^{(9)}(L_1)|, \tag{45}$$

where

$$S_c^{(9)}(L_1) = \sum_{1 \le h \le H} \left| \sum_{L < l \le L_1} e\left(h \Delta'_{(lt^2-1)d^{-2}} \right) \right| \tag{46}$$

for some number $L_1 \in (L, 2L]$. By (34), (37), (38), (46) and Lemma 2 with $q = 0$, we obtain

$$S_c^{(9)}(L_1) \ll \sum_{1 \le h \le H} \left(h^{\frac{1}{2}} L^{\frac{\gamma}{2}} t^{\gamma} + h^{-1} L^{1-\gamma} t^{-2\gamma} \right)$$

$$\ll L^{\frac{\gamma}{2}} H^{\frac{3}{2}} t^{\gamma} + L^{1-\gamma} t^{-2\gamma} \log H$$

$$\ll x^{\frac{1}{2}} H^{\frac{3}{2}} + x^{c-1} t^{-2} \log H. \tag{47}$$

Now, (45) and (47) imply

$$S_c^{(6)}(x) \ll x^{\frac{3-2c}{2}} H^{\frac{3}{2}} + t^{-2} \log H \ll x^{\frac{3-2c}{2}} H^{\frac{3}{2}}. \tag{48}$$

Taking into account (1), (24), (26), (30), (39), (40) and (48), we deduce

$$S_c^{(4)}(x) \ll \sum_{dt \le z} \left(x^c H^{-1} d^{-2} t^{-2} + x^{\frac{1}{2}} H^{\frac{1}{2}} + x^{\frac{3-2c}{2}} H^{\frac{3}{2}} \right)$$

$$\ll x^{\frac{2c+1}{4}} \sum_{dt \le z} (dt)^{-1} + x^{\frac{2c+3}{8}} \sum_{dt \le z} (dt)^{-\frac{1}{2}} + x^{\frac{9-2c}{8}} \sum_{dt \le z} (dt)^{-\frac{3}{2}}$$

$$\ll x^{\frac{2c+1}{4}} \sum_{n \le z} \tau(n) n^{-1} + x^{\frac{2c+3}{8}} \sum_{n \le z} \tau(n) n^{-\frac{1}{2}} + x^{\frac{9-2c}{8}}$$

$$\ll x^{\frac{2c+1}{4}+\varepsilon} \sum_{n \le z} n^{-1} + x^{\frac{2c+3}{8}+\varepsilon} \sum_{n \le z} n^{-\frac{1}{2}} + x^{\frac{9-2c}{8}}$$

$$\ll x^{\frac{2c+1}{4}+\varepsilon} + x^{\frac{2c+3}{8}+\varepsilon} z^{\frac{1}{2}} + x^{\frac{9-2c}{8}} \ll x^{\frac{9-2c}{8}}. \tag{49}$$

5.3. *Lower bound for $S_c^{(1)}(x)$*

Bearing in mind (14), (23) and (49), we derive

$$S_c^{(1)}(x) \gg x. \tag{50}$$

6. Upper Bound for $S_c^{(2)}(x)$

Splitting the range of d and t into dyadic subintervals from (12), we obtain

$$S_c^{(2)}(x) \ll \sum_{\substack{dt>z \\ (d,t)=1}} \sum_{\substack{(\Delta_1^c \tan^\theta (\log \Delta_1)-1)d^{-2}<k<\Delta_2^c \tan^\theta (\log \Delta_2)d^{-2} \\ kd^2+1\equiv 0\,(t^2)}} 1$$

$$\ll (\log x)^2 \sum_{D\leq d<2D} \sum_{T\leq t<2T}$$

$$\times \sum_{\substack{(\Delta_1^c \tan^\theta (\log \Delta_1)-1)d^{-2}<k<\Delta_2^c \tan^\theta (\log \Delta_2)d^{-2} \\ kd^2+1\equiv 0\,(t^2)}} 1, \qquad (51)$$

where

$$\frac{1}{2} \leq D,T \leq \sqrt{\Delta_2^c \tan^\theta (\log \Delta_2) + 1}, \quad DT > \frac{z}{4}. \qquad (52)$$

On the one hand, (3), (4) and (51) give us

$$S_c^{(2)}(x) \ll x^\varepsilon \sum_{T\leq t<2T} \sum_{l\leq (\Delta_2^c \tan^\theta (\log \Delta_2)+1)T^{-2}}$$

$$\times \sum_{D\leq d<2D} \sum_{\substack{(\Delta_1^c \tan^\theta (\log \Delta_1)-1)d^{-2}<k<\Delta_2^c \tan^\theta (\log \Delta_2)d^{-2} \\ kd^2=lt^2-1}} 1$$

$$\ll x^\varepsilon \sum_{T\leq t<2T} \sum_{l\leq (\Delta_2^c \tan^\theta (\log \Delta_2)+1)T^{-2}} \tau(lt^2-1)$$

$$\ll x^\varepsilon \sum_{T\leq t<2T} \sum_{l\leq (\Delta_2^c \tan^\theta (\log \Delta_2)+1)T^{-2}} 1$$

$$\ll x^{c+\varepsilon}T^{-1}. \qquad (53)$$

On the other hand, (3), (4) and (51) imply

$$S_c^{(2)}(x) \ll x^\varepsilon \sum_{D\leq d<2D} \sum_{(\Delta_1^c \tan^\theta (\log \Delta_1)-1)d^{-2}<k<\Delta_2^c \tan^\theta (\log \Delta_2)d^{-2}}$$

$$\times \sum_{T\leq t<2T} \sum_{\substack{l\leq (\Delta_2^c \tan^\theta (\log \Delta_2)+1)T^{-2} \\ kd^2+1=lt^2}} 1$$

$$\ll x^\varepsilon \sum_{D\leq d<2D} \sum_{k\leq \Delta_2^c \tan^\theta (\log \Delta_2)D^{-2}} \tau(kd^2+1)$$

$$\ll x^\varepsilon \sum_{D \le d < 2D} \sum_{k \le \Delta_2^c \tan^\theta (\log \Delta_2) D^{-2}} 1$$

$$\ll x^{c+\varepsilon} D^{-1}. \tag{54}$$

Now, (1) and (52)–(54) yield

$$S_c^{(2)}(x) \ll x^{c+\varepsilon} z^{-\frac{1}{2}} = x^{\frac{6c+1}{8}+\varepsilon}. \tag{55}$$

7.   The End of the Proof

Summarizing (11), (50) and (55), we establish the lower bound (8). This completes the proof of Theorem 1.

References

[1] X. Cao and W. Zhai (1998). The distribution of square-free numbers of the form $[n^c]$. *J. Théor. Nombres Bordeaux*, **10**, 287–299.

[2] X. Cao and W. Zhai (2008). The distribution of square-free numbers of the form $[n^c]$, II. *Acta Math. Sinica (Chin. Ser.)*, **51**, 1187–1194.

[3] L. Carlitz (1932). On a problem in additive arithmetic II. *Quart. J. Math.*, **3**, 273–290.

[4] S. I. Dimitrov, Prime numbers of the form $[n^c \tan^\theta (\log n)]$, arXiv:2110.11687.

[5] S. W. Graham and G. Kolesnik (1991). *Van der Corput's Method of Exponential Sums* (Cambridge University Press, New York).

[6] D. R. Heath-Brown (1984). The square-sieve and consecutive square-free numbers. *Math. Ann.*, **266**, 251–259.

[7] L. Mirsky (1949). On the frequency of pairs of square-free numbers with a given difference. *Bull. Amer. Math. Soc.*, **55**, 936–939.

[8] T. Reuss (2015). *The Determinant Method and Applications*, Thesis, (University of Oxford).

[9] G. J. Rieger (1978). Remark on a chapter of Stux concerning square-free numbers in non-linear sequences. *Pacific J. Math.*, **78**, 241–242.

[10] I. E. Stux, Distribution of square-free integers in non-linear sequences. *Pacific J. Math.*, **59**, 577–584.

[11] J. D. Vaaler (1985). Some extremal problems in Fourier analysis. *Bull. Amer. Math. Soc.*, **12**, 183–216.

Chapter 4

Inequalities for (m, M)-Ψ-Convex Functions with Applications to Operator Noncommutative Perspectives*

Silvestru Sever Dragomir

Departure of Mathematics, College of Engineering & Science,
Victoria University,
Melbourne City, MC 8001, Australia
School of Computer Science & Applied Mathematics,
DST-NRF Centre of Excellence in the
Mathematical and Statistical Sciences,
University of the Witwatersrand,
Private Bag 3, Johannesburg 2050, South Africa

sever.dragomir@vu.edu.au

In this chapter, we obtain some inequalities for (m, M)-Ψ-convex functions and apply them for operator noncommutative perspectives related to convex functions. Particular cases for weighted operator geometric mean and relative operator entropy are also given.

---

*This chapter is dedicated to my granddaughters Audrey and Sienna.

1.   Introduction

Assume that the function $\Psi : J \subseteq \mathbb{R} \to \mathbb{R}$ (J is an interval) is convex on J and $m \in \mathbb{R}$. We shall say that the function $\Phi : J \to \mathbb{R}$ is m-Ψ-*lower convex* if $\Phi - m\Psi$ is a convex function on J. We may introduce the class of functions [2]

$$\mathcal{L}(J, m, \Psi) := \{\Phi : J \to \mathbb{R} | \Phi - m\Psi \text{ is convex on } J\}. \qquad (1)$$

Similarly, for $M \in \mathbb{R}$ and Ψ as above, we can introduce the class of M-Ψ-*upper convex* functions by Ref. [2]

$$\mathcal{U}(J, M, \Psi) := \{\Phi : J \to \mathbb{R} | M\Psi - \Phi \text{ is convex on } J\}. \qquad (2)$$

The intersection of these two classes will be called the class of (m, M)-Ψ-*convex functions* and will be denoted by Ref. [2]

$$\mathcal{B}(J, m, M, \Psi) := \mathcal{L}(J, m, \Psi) \cap \mathcal{U}(J, M, \Psi). \qquad (3)$$

If $\Phi \in \mathcal{B}(J, m, M, \Psi)$, then $\Phi - m\Psi$ and $M\Psi - \Phi$ are convex and then $(\Phi - m\Psi) + (M\Psi - \Phi)$ is also convex, which shows that $(M - m)\Psi$ is convex, implying that $M \geq m$ (as Ψ is assumed not to be the trivial convex function $\Psi(t) = 0$, $t \in J$).

The above concepts may be introduced in the general case of a convex subset in a real linear space, but we do not consider this extension here.

In Ref. [14], Dragomir and Ionescu introduced the concept of *g-convex dominated functions* for a function $f : J \to \mathbb{R}$. We recall this, by saying, for a given convex function $g : J \to \mathbb{R}$, the function $f : J \to \mathbb{R}$ is *g-convex dominated* iff $g + f$ and $g - f$ are convex functions on J. In Ref. [14], the authors pointed out a number of inequalities for convex dominated functions related to the Jensens, Fuch, Pečarić, Barlow–Proschan and Vasić–Mijalković results. For more related results, see [4, 6, 8, 9, 16, 22, 23, 31].

We observe that the concept of g-convex dominated functions can be obtained as a particular case from (m, M)-Ψ-convex functions by choosing $m = -1$, $M = 1$ and $\Psi = g$.

The following lemma holds [2].

Lemma 1. *Let $\Psi, \Phi : J \subseteq \mathbb{R} \to \mathbb{R}$ be differentiable functions on $\mathring{J}$, the interior of J and Ψ is a convex function on J.*

(i) *For $m \in \mathbb{R}$, the function $\Phi \in \mathcal{L}\left(\mathring{J}, m, \Psi\right)$ iff*

$$m \left[\Psi\left(t\right) - \Psi\left(s\right) - \Psi'(s)(t - s)\right]$$
$$\leq \Phi\left(t\right) - \Phi\left(s\right) - \Phi'\left(s\right)\left(t - s\right) \tag{4}$$

for all $t, s \in \mathring{J}$.

(ii) *For $M \in \mathbb{R}$, the function $\Phi \in \mathcal{U}\left(\mathring{J}, M, \Psi\right)$ iff*

$$\Phi\left(t\right) - \Phi\left(s\right) - \Phi'\left(s\right)\left(t - s\right) \leq M \left[\Psi\left(t\right) - \Psi\left(s\right) - \Psi'\left(s\right)\left(t - s\right)\right] \tag{5}$$

for all $t, s \in \mathring{J}$.

(iii) *For $M, m \in \mathbb{R}$ with $M \geq m$, the function $\Phi \in \mathcal{B}\left(\mathring{J}, m, M, \Psi\right)$ iff both (4) and (5) hold.*

Another elementary fact for twice differentiable functions also holds [2].

Lemma 2. *Let $\Psi, \Phi : J \subseteq \mathbb{R} \to \mathbb{R}$ be twice differentiable on $\mathring{J}$ and Ψ is convex on J.*

(i) *For $m \in \mathbb{R}$, the function $\Phi \in \mathcal{L}\left(\mathring{J}, m, \Psi\right)$ iff*

$$m\Psi''\left(t\right) \leq \Phi''\left(t\right) \quad \text{for all } t \in \mathring{J}. \tag{6}$$

(ii) *For $M \in \mathbb{R}$, the function $\Phi \in \mathcal{U}\left(\mathring{J}, M, \Psi\right)$ iff*

$$\Phi''\left(t\right) \leq M\Psi''\left(t\right) \quad \text{for all } t \in \mathring{J}. \tag{7}$$

(iii) *For $M, m \in \mathbb{R}$ with $M \geq m$, the function $\Phi \in \mathcal{B}\left(\mathring{J}, m, M, \Psi\right)$ iff both (6) and (7) hold.*

For various inequalities concerning these classes of function, see the survey chapter [5].

Let Φ be a continuous function defined on the interval J of real numbers, B a self-adjoint operator on the Hilbert space H and A a positive invertible operator on H. Assume that the spectrum $\mathrm{Sp}\left(A^{-1/2}BA^{-1/2}\right) \subset \mathring{J}$. Then by using the continuous functional calculus, we can define the noncommutative *perspective* $\mathcal{P}_\Phi(B, A)$ by setting

$$\mathcal{P}_\Phi(B, A) := A^{1/2}\Phi\left(A^{-1/2}BA^{-1/2}\right)A^{1/2}.$$

If A and B are commutative, then

$$\mathcal{P}_\Phi(B, A) = A\Phi\left(BA^{-1}\right)$$

provided $\mathrm{Sp}\left(BA^{-1}\right) \subset \mathring{J}$.

It is known that (see Refs. [17–19]) if Φ is an *operator convex function* defined in the positive half-line, then the mapping

$$(B, A) \to \mathcal{P}_\Phi(B, A)$$

defined in pairs of positive definite operators is convex.

In a recent chapter, [10] we established the following reverse inequality for the perspective $\mathcal{P}_\Phi(B, A)$.

Let $\Phi : [a, b] \to \mathbb{R}$ be a *convex function* on the real interval $[a, b]$, A a positive invertible operator and B a self-adjoint operator such that

$$aA \leq B \leq bA, \tag{8}$$

then we have

$$0 \leq \frac{1}{b-a}\left[\Phi(a)(bA - B) + \Phi(b)(B - aA)\right] - \mathcal{P}_\Phi(B, A)$$

$$\leq \frac{\Phi'_-(b) - \Phi'_+(a)}{b-a}\left(bA^{1/2} - BA^{-1/2}\right)\left(A^{-1/2}B - aA^{1/2}\right)$$

$$\leq \frac{1}{4}(b-a)\left[\Phi'_-(b) - \Phi'_+(a)\right]A. \tag{9}$$

Let $\Phi : J \subset \mathbb{R} \to \mathbb{R}$ be a twice differentiable function on the interval $\mathring{J}$, the interior of J. Suppose that there exists the constants d, D such that

$$d \leq \Phi''(t) \leq D \quad \text{for any } t \in \mathring{J}. \tag{10}$$

If A is a positive invertible operator and B a self-adjoint operator such that the condition (8) is valid with $[a, b] \subset \mathring{J}$, then we have the following result as well [11]:

$$\frac{1}{2}d\left(bA^{1/2} - BA^{-1/2}\right)\left(A^{-1/2}B - aA^{1/2}\right)$$

$$\leq \frac{1}{b-a}\left[\Phi\left(a\right)\left(bA - B\right) + \Phi\left(b\right)\left(B - aA\right)\right] - \mathcal{P}_{\Phi}\left(B, A\right)$$

$$\leq \frac{1}{2}D\left(bA^{1/2} - BA^{-1/2}\right)\left(A^{-1/2}B - aA^{1/2}\right). \tag{11}$$

If $d > 0$, then the first inequality in (11) is better than the same inequality in (9).

Motivated by the above results, in this chapter, we obtain some inequalities for (m, M)-Ψ-convex functions and apply them for operator noncommutative perspectives related to convex functions. Particular cases for weighted operator geometric mean and relative operator entropy are also given.

2. Scalar Inequalities for (m, M)-Ψ-Convex Functions

We have the following simple fact that has several particular cases of interest for integrals and special means.

Proposition 1. *Assume that the function* $\Psi : J \subseteq \mathbb{R} \to \mathbb{R}$ *is convex on* J *and* $M, m \in \mathbb{R}$ *with* $M \geq m$. *Then the function* $\Phi : J \to \mathbb{R}$ *belongs to* $\mathcal{B}\left(J, m, M, \Psi\right)$ *if and only if it satisfies the double inequality*

$$m\left[(1 - \nu)\Psi\left(a\right) + \nu\Psi\left(b\right) - \Psi\left((1 - \nu)a + \nu b\right)\right]$$

$$\leq (1 - \nu)\Phi\left(a\right) + \nu\Phi\left(b\right) - \Phi\left((1 - \nu)a + \nu b\right)$$

$$\leq M\left[(1 - \nu)\Psi\left(a\right) + \nu\Psi\left(b\right) - \Psi\left((1 - \nu)a + \nu b\right)\right] \tag{12}$$

for any $a, b \in J$ *and any* $\nu \in [0, 1]$.

Proof. We have that $\Phi \in \mathcal{B}\left(J, m, M, \Psi\right)$ iff $\Phi - m\Psi$ and $M\Psi - \Phi$ and by the definition of convexity, this is equivalent to (12). □

Corollary 1. *Assume that the function* $\Psi : J \subseteq \mathbb{R} \to \mathbb{R}$ *is convex on* J *and* $M, m \in \mathbb{R}$ *with* $M \geq m$. *If the function* $\Phi : J \to \mathbb{R}$ *belongs to* $\mathcal{B}(J, m, M, \Psi)$, *then for any* $a, b \in J$ *with* $a < b$, *we have the inequalities*

$$m \left[\frac{(b-t)\,\Psi(a) + (t-a)\,\Psi(b)}{b-a} - \Psi(t) \right]$$

$$\leq \frac{(b-t)\,\Phi(a) + (t-a)\,\Phi(b)}{b-a} - \Phi(t)$$

$$\leq M \left[\frac{(b-t)\,\Psi(a) + (t-a)\,\Psi(b)}{b-a} - \Psi(t) \right] \tag{13}$$

and

$$m \left[\frac{(t-a)\,\Psi(a) + (b-t)\,\Psi(b)}{b-a} - \Psi(a+b-t) \right]$$

$$\leq \frac{(t-a)\,\Phi(a) + (b-t)\,\Phi(b)}{b-a} - \Phi(a+b-t)$$

$$\leq M \left[\frac{(b-t)\,\Psi(a) + (t-a)\,\Psi(b)}{b-a} - \Psi(a+b-t) \right] \tag{14}$$

for any $t \in [a, b]$.

In particular, we have

$$m \left[\frac{\Psi(a) + \Psi(b)}{2} - \Psi\left(\frac{a+b}{2}\right) \right]$$

$$\leq \frac{\Phi(a) + \Phi(b)}{2} - \Phi\left(\frac{a+b}{2}\right)$$

$$\leq M \left[\frac{\Psi(a) + \Psi(b)}{2} - \Psi\left(\frac{a+b}{2}\right) \right] \tag{15}$$

for any $a, b \in J$.

Proof.   The inequality (13) follows by (12) on taking $\nu = \frac{t-a}{b-a} \in [0,1]$ while (14) follows by (12) on taking $\nu = \frac{b-t}{b-a} \in [0,1]$.   □

Remark 1. By adding the inequalities (13) and (14) and dividing by 2, we get

$$m\left[\frac{\Psi(a) + \Psi(b)}{2} - \frac{\Psi(t) + \Psi(a + b - t)}{2}\right]$$

$$\leq \frac{\Phi(a) + \Phi(b)}{2} - \frac{\Phi(t) + \Phi(a + b - t)}{2}$$

$$\leq M\left[\frac{\Psi(a) + \Psi(b)}{2} - \frac{\Psi(t) + \Psi(a + b - t)}{2}\right] \qquad (16)$$

for any $t \in [a, b]$.

If we replace in (15) a by t and b by $a + b - t$, we get

$$m\left[\frac{\Psi(t) + \Psi(a + b - t)}{2} - \Psi\left(\frac{a + b}{2}\right)\right]$$

$$\leq \frac{\Phi(t) + \Phi(a + b - t)}{2} - \Phi\left(\frac{a + b}{2}\right)$$

$$\leq M\left[\frac{\Psi(t) + \Psi(a + b - t)}{2} - \Psi\left(\frac{a + b}{2}\right)\right] \qquad (17)$$

for any $t \in [a, b]$.

We have the following Hermite–Hadamard-type result.

Theorem 1. *Assume that the function* $\Psi : J \subseteq \mathbb{R} \to \mathbb{R}$ *is convex on* J *and* $M, m \in \mathbb{R}$ *with* $M \geq m$. *If the function* $\Phi : J \to \mathbb{R}$ *belongs to* $\mathcal{B}(J, m, M, \Psi)$, *then for any* $a, b \in J$, $b \neq a$, *we have the inequalities*

$$m\left[\frac{\Psi(a) + \Psi(b)}{2} - \frac{1}{b - a}\int_a^b \Psi(t)\,dt\right]$$

$$\leq \frac{\Phi(a) + \Phi(b)}{2} - \frac{1}{b - a}\int_a^b \Phi(t)\,dt$$

$$\leq M\left[\frac{\Psi(a) + \Psi(b)}{2} - \frac{1}{b - a}\int_a^b \Psi(t)\,dt\right] \qquad (18)$$

and

$$m \left[\frac{1}{b-a} \int_a^b \Psi(t)\, dt - \Psi\left(\frac{a+b}{2}\right) \right]$$

$$\leq \frac{1}{b-a} \int_a^b \Phi(t)\, dt - \Phi\left(\frac{a+b}{2}\right)$$

$$\leq M \left[\frac{1}{b-a} \int_a^b \Psi(t)\, dt - \Psi\left(\frac{a+b}{2}\right) \right]. \tag{19}$$

Proof. If we integrate the inequality (12) over $\nu \in [0,1]$ and use the change of variable $t = (1-\nu)\,a + \nu b$, $\nu \in [0,1]$, we get the desired result (18).

From the inequality (12), we have

$$m \left[\frac{\Psi(s) + \Psi(t)}{2} - \Psi\left(\frac{s+t}{2}\right) \right]$$

$$\leq \left[\frac{\Phi(s) + \Phi(t)}{2} - \Phi\left(\frac{s+t}{2}\right) \right]$$

$$\leq M \left[\frac{\Psi(s) + \Psi(t)}{2} - \Psi\left(\frac{s+t}{2}\right) \right] \tag{20}$$

for any $s, t \in J$.

If we take in (20) $s = (1-\nu)\,a + \nu b$ and $t = (1-\nu)\,b + \nu a$ with $a, b \in J$ and $\nu \in [0,1]$, then we get

$$m \left[\frac{\Psi((1-\nu)\,a + \nu b) + \Psi((1-\nu)\,b + \nu a)}{2} - \Psi\left(\frac{a+b}{2}\right) \right]$$

$$\leq \left[\frac{\Phi((1-\nu)\,a + \nu b) + \Phi((1-\nu)\,b + \nu a)}{2} - \Phi\left(\frac{a+b}{2}\right) \right]$$

$$\leq M \left[\frac{\Psi((1-\nu)\,a + \nu b) + \Psi((1-\nu)\,b + \nu a)}{2} - \Psi\left(\frac{a+b}{2}\right) \right] \tag{21}$$

for $a, b \in J$ and $\nu \in [0,1]$.

If we integrate the inequality (21) over $\nu \in [0, 1]$ and use the fact that

$$\int_0^1 \Psi\left((1-\nu)\, a + \nu b\right) d\nu = \int_0^1 \Psi\left((1-\nu)\, b + \nu a\right) d\nu$$

$$= \frac{1}{b-a} \int_a^b f(t)\, dt,$$

then we get the desired result (19).   □

Theorem 2. *Let* $\Psi, \Phi : J \subseteq \mathbb{R} \to \mathbb{R}$ *be differentiable functions on* $\mathring{J}$ *and* Ψ *is a convex function on* J. *If the function* $\Phi : J \to \mathbb{R}$ *belongs to* $\mathcal{B}(J, m, M, \Psi)$, *then for any* $a, b \in J$, $b \neq a$, *we have the inequalities*

$$m\left[\frac{1}{b-a} \int_a^b \Psi(t)\, dt - \Psi(s) - \Psi'(s)\left(\frac{a+b}{2} - s\right)\right]$$

$$\leq \frac{1}{b-a} \int_a^b \Phi(t)\, dt - \Phi(s) - \Phi'(s)\left(\frac{a+b}{2} - s\right)$$

$$\leq M\left[\frac{1}{b-a} \int_a^b \Psi(t)\, dt - \Psi(s) - \Psi'(s)\left(\frac{a+b}{2} - s\right)\right] \quad (22)$$

for any $s \in \mathring{J}$ *and*

$$m\left[\frac{1}{2}\left(\Psi(t) + \frac{\Psi(b)\,(b-t) + \Psi(a)\,(t-a)}{b-a}\right)\right.$$

$$\left. - \frac{1}{b-a} \int_a^b \Psi(s)\, ds\right]$$

$$\leq \frac{1}{2}\left(\Phi(t) + \frac{\Phi(b)\,(b-t) + \Phi(a)\,(t-a)}{b-a}\right)$$

$$- \frac{1}{b-a} \int_a^b \Phi(s)\, ds$$

$$\leq M\left[\frac{1}{2}\left(\Psi(t) + \frac{\Psi(b)\,(b-t) + \Psi(a)\,(t-a)}{b-a}\right)\right.$$

$$\left. - \frac{1}{b-a} \int_a^b \Psi(s)\, ds\right] \quad (23)$$

for any $t \in J$.

Proof. Since $\Phi \in \mathcal{B}(J, m, M, \Psi)$, then by Lemma 1, we have

$$m\left[\Psi(t) - \Psi(s) - \Psi'(s)(t-s)\right]$$
$$\leq \Phi(t) - \Phi(s) - \Phi'(s)(t-s)$$
$$\leq M\left[\Psi(t) - \Psi(s) - \Psi'(s)(t-s)\right] \tag{24}$$

for all $s \in \mathring{J}$ and $t \in J$.

Let $a, b \in J$, $b \neq a$. The integral mean $\frac{1}{b-a}\int_a^b$ is a linear positive functional.

Now, by taking the integral mean over t in (24), we get

$$m\left[\frac{1}{b-a}\int_a^b \Psi(t)\,dt - \Psi(s) - \Psi'(s)\left(\frac{1}{b-a}\int_a^b t\,dt - s\right)\right]$$
$$\leq \frac{1}{b-a}\int_a^b \Phi(t)\,dt - \Phi(s) - \Phi'(s)\left(\frac{1}{b-a}\int_a^b t\,dt - s\right)$$
$$\leq M\left[\frac{1}{b-a}\int_a^b \Psi(t)\,dt - \Psi(s) - \Psi'(s)\left(\frac{1}{b-a}\int_a^b t\,dt - s\right)\right]$$

for any $s \in \mathring{J}$ that is equivalent to (22).

Now, if we take the integral mean in (24) over s, we get

$$m\left[\Psi(t) - \frac{1}{b-a}\int_a^b \Psi(s)\,ds - \frac{1}{b-a}\int_a^b \Psi'(s)(t-s)\,ds\right]$$
$$\leq \Phi(t) - \frac{1}{b-a}\int_a^b \Phi(s)\,ds - \frac{1}{b-a}\int_a^b \Phi'(s)(t-s)\,ds$$
$$\leq M\left[\Psi(t) - \frac{1}{b-a}\int_a^b \Psi(s)\,ds - \frac{1}{b-a}\int_a^b \Psi'(s)(t-s)\,ds\right] \tag{25}$$

for any $t \in J$.

Observe that, integrating by parts, we have

$$\int_a^b \Psi'(s)(t-s)\,ds = \Psi(s)(t-s)\big|_a^b + \int_a^b \Psi(s)\,ds$$
$$= \Psi(b)(t-b) - \Psi(a)(t-a) + \int_a^b \Psi(s)\,ds$$
$$= -\Psi(b)(b-t) - \Psi(a)(t-a) + \int_a^b \Psi(s)\,ds,$$

which implies that

$$\Psi(t) - \frac{1}{b-a} \int_a^b \Psi(s)\, ds - \frac{1}{b-a} \int_a^b \Psi'(s)(t-s)\, ds$$

$$= \Psi(t) + \frac{\Psi(b)(b-t) + \Psi(a)(t-a)}{b-a} - \frac{2}{b-a} \int_a^b \Psi(s)\, ds$$

and a similar equality for Φ.

By (25), we get then the desired result (23). □

Remark 2. If we take $s = \frac{a+b}{2}$ in (22), then we recapture the inequality (19). The same choice for t in (23) produces the inequality

$$m\left[\frac{1}{2}\left(\Psi\left(\frac{a+b}{2}\right) + \frac{\Psi(a) + \Psi(b)}{2}\right) - \frac{1}{b-a} \int_a^b \Psi(s)\, ds\right]$$

$$\leq \frac{1}{2}\left(\Phi\left(\frac{a+b}{2}\right) + \frac{\Phi(b) + \Phi(a)}{2}\right) - \frac{1}{b-a} \int_a^b \Phi(s)\, ds$$

$$\leq M\left[\frac{1}{2}\left(\Psi\left(\frac{a+b}{2}\right) + \frac{\Psi(a) + \Psi(b)}{2}\right) - \frac{1}{b-a} \int_a^b \Psi(s)\, ds\right].$$

$$(26)$$

Remark 3. If the function $\Phi \in \mathcal{B}(J, m, M, \Psi)$, then $\Phi = m\Psi + g$ with Ψ and g are convex functions, implying that Φ is differentiable everywhere except a countable number of points in J and the inequality (24) holds for almost every $s \in \mathring{J}$ and every $t \in J$. Making use of a similar argument as above, we conclude that the inequality (23) holds without the differentiability assumption.

3. Inequalities for Perspectives

We have the following.

Theorem 3. *Let* $\Phi : J \to \mathbb{R}$ *be a convex function on the interval of real numbers* J, M, $m \in \mathbb{R}$ *with* $M \geq m$ *and the function* $\Phi : J \to \mathbb{R}$ *that belongs to* $\mathcal{B}(J, m, M, \Psi)$. *If* A *is a positive invertible operator and* B *is a self-adjoint operator such that*

$$aA \leq B \leq bA \tag{27}$$

with $[a, b] \subset \mathring{J}$ *for some real numbers* $a < b$, *then we have the inequalities*

$$m \left[\frac{1}{b-a} \left[\Psi(a)(bA - B) + \Psi(b)(B - aA) \right] - \mathcal{P}_\Psi(B, A) \right]$$

$$\leq \frac{1}{b-a} \left[\Phi(a)(bA - B) + \Phi(b)(B - aA) \right] - \mathcal{P}_\Phi(B, A)$$

$$\leq M \left[\frac{1}{b-a} \left[\Psi(a)(bA - B) + \Psi(b)(B - aA) \right] \right.$$

$$\left. - \mathcal{P}_\Psi(B, A) \right] \tag{28}$$

and

$$m \left[\frac{1}{b-a} \Psi(a)(B - aA) + \Psi(b)(bA - B) - \mathcal{P}_{\Psi(a+b-\cdot)}(B, A) \right]$$

$$\leq \frac{1}{b-a} \Phi(a)(B - aA) + \Phi(b)(bA - B) - \mathcal{P}_{\Phi(a+b-\cdot)}(B, A)$$

$$\leq M \left[\frac{1}{b-a} \Psi(a)(B - aA) + \Psi(b)(bA - B) \right.$$

$$\left. - \mathcal{P}_{\Psi(a+b-\cdot)}(B, A) \right], \tag{29}$$

where

$$\mathcal{P}_{\Phi(a+b-\cdot)}(B, A) = A^{1/2} \Phi \left(A^{-1/2} \left[(a + b)A - B \right] A^{-1/2} \right) A^{1/2}$$

and a similar expression for Ψ.

   Moreover, we have

$$m \left[\frac{\Psi(a) + \Psi(b)}{2} A - \frac{\mathcal{P}_\Psi(B, A) + \mathcal{P}_{\Psi(a+b-\cdot)}(B, A)}{2} \right]$$

$$\leq \frac{\Phi(a) + \Phi(b)}{2} A - \frac{\mathcal{P}_\Phi(B, A) + \mathcal{P}_{\Phi(a+b-\cdot)}(B, A)}{2}$$

$$\leq M \left[\frac{\Psi(a) + \Psi(b)}{2} A - \frac{\mathcal{P}_\Psi(B, A) + \mathcal{P}_{\Psi(a+b-\cdot)}(B, A)}{2} \right]. \tag{30}$$

Proof. Using the continuous functional calculus, for any self-adjoint operator X with $\mathrm{Sp}\,(X) \subseteq [a, b]$, we have from (13) that

$$m \left[\frac{\Psi\,(a)\,(bI - X) + \Psi\,(b)\,(X - aI)}{b - a} - \Psi\,(X) \right]$$

$$\leq \frac{\Phi\,(a)\,(bI - X) + \Phi\,(b)\,(X - aI)}{b - a} - \Phi\,(X)$$

$$\leq M \left[\frac{\Psi\,(a)\,(bI - X) + \Psi\,(b)\,(X - aI)}{b - a} - \Psi\,(X) \right] \qquad (31)$$

in the operator order.

If (27) holds, then by multiplying both sides with $A^{-1/2}$, we get

$$aI \leq A^{-1/2}BA^{-1/2} \leq bI,$$

and by writing the inequality (31) for $X = A^{-1/2}BA^{-1/2}$, we get

$$m \left[\frac{1}{b - a} \left[\Psi\,(a) \left(bI - A^{-1/2}BA^{-1/2} \right) \right. \right.$$

$$\left. \left. + \Psi\,(b) \left(A^{-1/2}BA^{-1/2} - aI \right) \right] - \Psi\left(A^{-1/2}BA^{-1/2} \right) \right]$$

$$\leq \frac{1}{b - a} \left[\Phi\,(a) \left(bI - A^{-1/2}BA^{-1/2} \right) \right.$$

$$\left. + \Phi\,(b) \left(A^{-1/2}BA^{-1/2} - aI \right) \right] - \Phi\left(A^{-1/2}BA^{-1/2} \right)$$

$$\leq M \left[\frac{1}{b - a} \left[\Psi\,(a) \left(bI - A^{-1/2}BA^{-1/2} \right) \right. \right.$$

$$\left. \left. + \Psi(b) \left(A^{-1/2}BA^{-1/2} - aI \right) \right] - \Psi\left(A^{-1/2}BA^{-1/2} \right) \right]$$

that can be rewritten as

$$m \left[\frac{1}{b - a} \left[\Psi\,(a)\,A^{-1/2}\,(bA - B)\,A^{-1/2} + \Psi\,(b)\,A^{-1/2} \right. \right.$$

$$\left. \left. \times\,(B - aA)\,A^{-1/2} \right] - \Psi\left(A^{-1/2}BA^{-1/2} \right) \right]$$

$$\leq \frac{1}{b-a} \left[\Phi\left(a\right) A^{-1/2} \left(bA - B\right) A^{-1/2} + \Phi\left(b\right) A^{-1/2} \right.$$

$$\left. \times \left(B - aA\right) A^{-1/2} \right] - \Phi\left(A^{-1/2} B A^{-1/2}\right)$$

$$\leq M \left[\frac{1}{b-a} \Psi\left(a\right) A^{-1/2} \left(bA - B\right) A^{-1/2} + \Psi\left(b\right) A^{-1/2} \right.$$

$$\left. \times \left(B - aA\right) A^{-1/2} - \Psi\left(A^{-1/2} B A^{-1/2}\right) \right]. \tag{32}$$

If we multiply (32) both sides with $A^{1/2}$, we get the desired result (28).

The inequality (29) follows in a similar way from (29) and we omit the details.

If we add (28) and (29) and divide by 2, we get (30). ☐

Theorem 4. *Let* $\Psi, \Phi : J \subseteq \mathbb{R} \to \mathbb{R}$ *be continuously differentiable functions on* $\mathring{J}$ *and* Ψ *is a convex function on* J. *If the function* $\Phi : J \to \mathbb{R}$ *belongs to* $\mathcal{B}\left(J, m, M, \Psi\right)$, A *is a positive invertible operator and* B *a self-adjoint operator such that the condition* (27) *is valid with* $[a, b] \subset \mathring{J}$ *for some real numbers* $a < b$, *then we have the inequalities*

$$m\left[\left(\frac{1}{b-a} \int_a^b \Psi\left(t\right) dt\right) A - \mathcal{P}_\Psi\left(B, A\right) - \mathcal{P}_{\Psi'(\cdot)\left(\frac{a+b}{2} - \cdot\right)}\left(B, A\right)\right]$$

$$\leq \left(\frac{1}{b-a} \int_a^b \Phi\left(t\right) dt\right) A - \mathcal{P}_\Phi\left(B, A\right) - \mathcal{P}_{\Phi'(\cdot)\left(\frac{a+b}{2} - \cdot\right)}\left(B, A\right)$$

$$\leq M\left[\left(\frac{1}{b-a} \int_a^b \Psi\left(t\right) dt\right) A - \mathcal{P}_\Psi\left(B, A\right) \right.$$

$$\left. - \mathcal{P}_{\Psi'(\cdot)\left(\frac{a+b}{2} - \cdot\right)}\left(B, A\right)\right], \tag{33}$$

where

$$\mathcal{P}_{\Phi'(\cdot)\left(\frac{a+b}{2} - \cdot\right)}\left(B, A\right) = A^{1/2} \Phi'\left(A^{-1/2} B A^{-1/2}\right) A^{-1/2} \left(\frac{a+b}{2} A - B\right)$$

and a similar expression for Ψ.

Proof. Using the continuous functional calculus, for any self-adjoint operator X with $\mathrm{Sp}(X) \subseteq [a, b]$, we have from (22) that

$$m \left[\left(\frac{1}{b-a} \int_a^b \Psi(t) \, dt \right) I - \Psi(X) - \Psi'(X) \left(\frac{a+b}{2} I - X \right) \right]$$

$$\leq \left(\frac{1}{b-a} \int_a^b \Phi(t) \, dt \right) I - \Phi(X) - \Phi'(X) \left(\frac{a+b}{2} I - X \right)$$

$$\leq M \left[\left(\frac{1}{b-a} \int_a^b \Psi(t) \, dt \right) I - \Psi(X) \right.$$

$$\left. - \Psi'(X) \left(\frac{a+b}{2} I - X \right) \right]. \tag{34}$$

If we write the inequality (34) for $X = A^{-1/2} B A^{-1/2}$, we get

$$m \left[\left(\frac{1}{b-a} \int_a^b \Psi(t) \, dt \right) I - \Psi\left(A^{-1/2} B A^{-1/2} \right) \right.$$

$$\left. - \Psi'\left(A^{-1/2} B A^{-1/2} \right) \left(\frac{a+b}{2} I - A^{-1/2} B A^{-1/2} \right) \right]$$

$$\leq \left(\frac{1}{b-a} \int_a^b \Phi(t) \, dt \right) I - \Phi\left(A^{-1/2} B A^{-1/2} \right)$$

$$- \Phi'\left(A^{-1/2} B A^{-1/2} \right) \left(\frac{a+b}{2} I - A^{-1/2} B A^{-1/2} \right)$$

$$\leq M \left[\left(\frac{1}{b-a} \int_a^b \Psi(t) \, dt \right) I - \Psi\left(A^{-1/2} B A^{-1/2} \right) \right.$$

$$\left. - \Psi'\left(A^{-1/2} B A^{-1/2} \right) \left(\frac{a+b}{2} I - A^{-1/2} B A^{-1/2} \right) \right]. \tag{35}$$

If we multiply (35) both sides with $A^{1/2}$, we get the desired result (33). $\qquad \square$

We also have the following:

Theorem 5. *Let* $\Phi : J \to \mathbb{R}$ *be a convex function on the interval of real numbers* J, M, $m \in \mathbb{R}$ *with* $M \geq m$ *and the function* $\Phi : J \to \mathbb{R}$ *that belongs to* $\mathcal{B}(J, m, M, \Psi)$. *If* A *is a positive invertible operator and* B *is a self-adjoint operator such that the condition* (27) *is valid with* $[a, b] \subset \mathring{J}$ *for some real numbers* $a < b$, *then we have the inequalities*

$$m \left[\frac{1}{2} \left(\mathcal{P}_\Psi (B, A) + \frac{1}{b-a} [\Psi (b) (bA - B) + \Psi (a) (B - aA)] \right) \right.$$

$$\left. - \left(\frac{1}{b-a} \int_a^b \Psi (s) \, ds \right) A \right]$$

$$\leq \frac{1}{2} \left(\mathcal{P}_\Phi (B, A) + \frac{1}{b-a} [\Phi (b) (bA - B) + \Phi (a) (B - aA)] \right)$$

$$- \left(\frac{1}{b-a} \int_a^b \Phi (s) \, ds \right) A$$

$$\leq M \left[\frac{1}{2} \left(\mathcal{P}_\Psi (B, A) + \frac{1}{b-a} [\Psi (b) (bA - B) \right. \right.$$

$$\left. \left. + \Psi (a) (B - aA)] \right) - \left(\frac{1}{b-a} \int_a^b \Psi (s) \, ds \right) A \right]. \qquad (36)$$

The proof follows in a similar way as above by employing the scalar inequality (23) and we omit the details.

4. Applications for Power Function

Let $p \in (-\infty, 0) \cup (1, \infty)$ and $\Phi : J \subset (0, \infty) \to \mathbb{R}$ be twice differentiable on $\mathring{J}$ and such that for some γ, Γ, we have

$$\gamma x^{p-2} \leq \Phi'' (x) \leq \Gamma x^{p-2} \quad \text{for any } x \in \mathring{J}. \qquad (37)$$

We observe that the functions $\Phi - \frac{\gamma}{p(p-1)} \ell^p$ and $\frac{\Gamma}{p(p-1)} \ell^p - \Phi$ where ℓ is the identity function, i.e. $\ell (t) = t$, are convex functions on J. Since $\Psi := \ell^p$ is also a convex function, it follows that $\Phi \in \mathcal{B} (J, m, M, \Psi)$ with $m = \frac{\gamma}{p(p-1)}$ and $M = \frac{\Gamma}{p(p-1)}$.

If we use the inequality (12), then we have the inequality

$$\frac{\gamma}{p(p-1)} [(1 - \nu) a^p + \nu b^p - ((1 - \nu) a + \nu b)^p]$$

$$\leq (1 - \nu) \Phi (a) + \nu \Phi (b) - \Phi ((1 - \nu) a + \nu b)$$

$$\leq \frac{\Gamma}{p(p-1)} [(1 - \nu) a^p + \nu b^p - ((1 - \nu) a + \nu b)^p] \qquad (38)$$

for any $a, b \in J$ and $\nu \in [0, 1]$.

If we take $p = 2$ in (38), then we get

$$\frac{\gamma}{2}(1 - \nu)\nu(b - a)^2 \le (1 - \nu)\Phi(a) + \nu\Phi(b)$$

$$-\Phi((1 - \nu)a + \nu b)$$

$$\le \frac{\Gamma}{2}(1 - \nu)\nu(b - a)^2 \qquad (39)$$

for any $a, b \in J$ and $\nu \in [0, 1]$, provided Φ is twice differentiable and

$$\gamma \le \Phi''(x) \le \Gamma \quad \text{for any } x \in \mathring{J}. \qquad (40)$$

We observe that inequality (39) is a particular case of inequality (2.7) [3], for $n = 2$, $p_1 = \nu$, $p_2 = 1 - \nu$ and when the inner product space H reduces to $\mathbb{R}$.

It has been obtained in this form recently in Ref. [1] by the interesting two variable functions analysis argument and in Ref. [7] by a convexity argument similar to the one from above.

If Φ is twice differentiable and

$$\gamma x^{-3} \le \Phi''(x) \le \Gamma x^{-3} \quad \text{for any } x \in \mathring{J}, \qquad (41)$$

then by taking $p = -1$ in (38), we get

$$\frac{\gamma}{2}\left[(1 - \nu)a^{-1} + \nu b^{-1} - ((1 - \nu)a + \nu b)^{-1}\right]$$

$$\le (1 - \nu)\Phi(a) + \nu\Phi(b) - \Phi((1 - \nu)a + \nu b)$$

$$\le \frac{\Gamma}{2}\left[(1 - \nu)a^{-1} + \nu b^{-1} - ((1 - \nu)a + \nu b)^{-1}\right] \qquad (42)$$

for any $a, b \in J$ and $\nu \in [0, 1]$.

Upon appropriate calculations, we have the following equivalent inequality:

$$\frac{\gamma}{2}(1 - \nu)\nu\frac{(b - a)^2}{ab[(1 - \nu)a + \nu b]}$$

$$\le (1 - \nu)\Phi(a) + \nu\Phi(b)$$

$$- \Phi((1 - \nu)a + \nu b)$$

$$\le \frac{\Gamma}{2}(1 - \nu)\nu\frac{(b - a)^2}{ab[(1 - \nu)a + \nu b]} \qquad (43)$$

for any $a, b \in J$ and $\nu \in [0, 1]$.

Since

$$\frac{\min\{a,b\}}{ab} = \frac{1}{\max\{a,b\}} \le \frac{1}{(1-\nu)a+\nu b} \le \frac{1}{\min\{a,b\}}$$

$$= \frac{\max\{a,b\}}{ab},$$

then by (43), we have the simpler, however coarser, inequality

$$\frac{\gamma}{2}(1-\nu)\nu\min\{a,b\}\left(\frac{1}{a}-\frac{1}{b}\right)^2$$

$$\le (1-\nu)\Phi(a)+\nu\Phi(b)-\Phi((1-\nu)a+\nu b)$$

$$\le \frac{\Gamma}{2}(1-\nu)\nu\max\{a,b\}\left(\frac{1}{a}-\frac{1}{b}\right)^2 \tag{44}$$

for any $a,b \in J$ and $\nu \in [0,1]$.

Let $q \in (0,1)$ and $\Phi : J \subset (0,\infty) \to \mathbb{R}$ be twice differentiable on $\mathring{J}$ and such that for some γ, Γ, we have

$$\gamma x^{q-2} \le \Phi''(x) \le \Gamma x^{q-2} \quad \text{for any } x \in \mathring{J}. \tag{45}$$

Since $\Psi := -\ell^q$ is a convex function and

$$\Phi - \frac{\gamma}{q(1-q)}(-\ell^q) = \Phi - \frac{\gamma}{q(q-1)}\ell^q$$

and

$$\frac{\Gamma}{q(1-q)}(-\ell^q) - \Phi = \frac{\Gamma}{q(q-1)}\ell^q - \Phi,$$

it follows that $\Phi \in \mathcal{B}(J,m,M,\Psi)$ with $m = \frac{\gamma}{q(1-q)}$ and $M = \frac{\Gamma}{q(q-1)}$.

By using the inequality (12), we have

$$\frac{\gamma}{q(1-q)}[((1-\nu)a+\nu b)^q - (1-\nu)a^q - \nu b^q]$$

$$\le (1-\nu)\Phi(a)+\nu\Phi(b)-\Phi((1-\nu)a+\nu b)$$

$$\le \frac{\Gamma}{q(1-q)}[((1-\nu)a+\nu b)^q - (1-\nu)a^q - \nu b^q] \tag{46}$$

for any $a,b \in J$ and $\nu \in [0,1]$.

If we take $q = \frac{1}{2}$, then we have

$$4\gamma \left[\sqrt{(1 - \nu) a + \nu b} - (1 - \nu) \sqrt{a} - \nu \sqrt{b} \right]$$

$$\leq (1 - \nu) \Phi (a) + \nu \Phi (b) - \Phi ((1 - \nu) a + \nu b)$$

$$\leq 4\Gamma \left[\sqrt{(1 - \nu) a + \nu b} - (1 - \nu) \sqrt{a} - \nu \sqrt{b} \right] \tag{47}$$

for any $a, b \in J$ and $\nu \in [0, 1]$.

For positive $x \neq y$ and $p \in \mathbb{R} \setminus \{-1, 0\}$, we define the p-*logarithmic mean* (*generalized logarithmic mean*) $L_p(x, y)$ by

$$L_p(x, y) := \left[\frac{y^{p+1} - x^{p+1}}{(p + 1)(y - x)} \right]^{1/p}.$$

In fact, the singularities at $p = -1$, 0 are removable and L_p can be defined for $p = -1$, 0 so as to make $L_p(x, y)$ a continuous function of p. In the limit as $p \to 0$, we obtain the *identric mean* $I(x, y)$, given by

$$I(x, y) := \frac{1}{e} \left(\frac{y^y}{x^x} \right)^{1/(y-x)}, \tag{48}$$

and in the case $p \to -1$, the *logarithmic mean* $L(x, y)$ is given by

$$L(x, y) := \frac{y - x}{\ln y - \ln x}.$$

In each case, we define the mean as x when $y = x$, which occurs as the limiting value of $L_p(x, y)$ for $y \to x$.

We define the arithmetic mean as $A(x, y) := \frac{x+y}{2}$, the geometric mean as $G(x, y) := \sqrt{xy}$, and the harmonic mean as $H(x, y) = A^{-1}(x^{-1}, y^{-1})$.

Let $p \in (-\infty, 0) \cup (1, \infty)$ and $\Phi : J \subset (0, \infty) \to \mathbb{R}$ be twice differentiable on $\mathring{J}$ and such that for some γ, Γ, we have the condition (37). Then by (18) and (19), we have

$$\frac{\gamma}{p(p - 1)} \left[A(a^p, b^p) - L_p^p(a, b) \right]$$

$$\leq \frac{\Phi(a) + \Phi(b)}{2} - \frac{1}{b - a} \int_a^b \Phi(t) \, dt$$

$$\leq \frac{\Gamma}{p(p - 1)} \left[A(a^p, b^p) - L_p^p(a, b) \right] \tag{49}$$

and

$$\frac{\gamma}{p(p-1)} \left[L_p^p(a,b) - A^p(a,b) \right]$$

$$\leq \frac{1}{b-a} \int_a^b \Phi(t)\, dt - \Phi\left(\frac{a+b}{2}\right)$$

$$\leq \frac{\Gamma}{p(p-1)} \left[L_p^p(a,b) - A^p(a,b) \right]. \tag{50}$$

From (26), we have

$$\frac{\gamma}{p(p-1)} \left[\frac{1}{2} \left(A^p(a,b) + A(a^p,b^p) \right) - L_p^p(a,b) \right]$$

$$\leq \frac{1}{2} \left(\Phi\left(\frac{a+b}{2}\right) + \frac{\Phi(b) + \Phi(a)}{2} \right) - \frac{1}{b-a} \int_a^b \Phi(s)\, ds$$

$$\leq \frac{\Gamma}{p(p-1)} \left[\frac{1}{2} \left(A^p(a,b) + A(a^p,b^p) \right) - L_p^p(a,b) \right]. \tag{51}$$

If we take $p = 2$ in (49)–(51), then we get

$$\frac{1}{12}\gamma(b-a)^2 \leq \frac{\Phi(a) + \Phi(b)}{2} - \frac{1}{b-a} \int_a^b \Phi(t)\, dt$$

$$\leq \frac{1}{12}\Gamma(b-a)^2, \tag{52}$$

$$\frac{1}{24}\gamma(b-a)^2 \leq \frac{1}{b-a} \int_a^b \Phi(t)\, dt - \Phi\left(\frac{a+b}{2}\right)$$

$$\leq \frac{1}{24}\Gamma(b-a)^2. \tag{53}$$

and

$$\frac{1}{48}\gamma(b-a)^2 \leq \frac{1}{2} \left(\Phi\left(\frac{a+b}{2}\right) + \frac{\Phi(b) + \Phi(a)}{2} \right)$$

$$- \frac{1}{b-a} \int_a^b \Phi(s)\, ds \leq \frac{1}{48}\Gamma(b-a)^2, \tag{54}$$

provided Φ is twice differentiable and satisfies condition (40), see also Ref. [5].

If Φ is twice differentiable and the condition (41) is satisfied, then by taking $p = -1$ in (49)–(51), we get

$$\frac{1}{2}\gamma \frac{L(a,b) - H(a,b)}{L(a,b) H(a,b)} \leq \frac{\Phi(a) + \Phi(b)}{2} - \frac{1}{b-a} \int_a^b \Phi(t)\, dt$$

$$\leq \frac{1}{2}\Gamma \frac{L(a,b) - H(a,b)}{L(a,b) H(a,b)}, \tag{55}$$

$$\frac{1}{2}\gamma \frac{A(a,b) - L(a,b)}{A(a,b) L(a,b)} \leq \frac{1}{b-a} \int_a^b \Phi(t)\, dt - \Phi\left(\frac{a+b}{2}\right)$$

$$\leq \frac{1}{2}\Gamma \frac{A(a,b) - L(a,b)}{A(a,b) L(a,b)} \tag{56}$$

and

$$\frac{1}{4}\gamma \frac{L(a,b)\left[A(a,b) + H(a,b)\right] - A(a,b) H(a,b)}{A(a,b) L(a,b) H(a,b)}$$

$$\leq \frac{1}{2}\left(\Phi\left(\frac{a+b}{2}\right) + \frac{\Phi(b) + \Phi(a)}{2}\right) - \frac{1}{b-a} \int_a^b \Phi(s)\, ds$$

$$\leq \frac{1}{4}\Gamma \frac{L(a,b)\left[A(a,b) + H(a,b)\right] - A(a,b) H(a,b)}{A(a,b) L(a,b) H(a,b)}. \tag{57}$$

Assume that A, B are positive invertible operators on a complex Hilbert space $(H, \langle \cdot, \cdot \rangle)$. We use the following notations for operators [28]:

$$A \nabla_\nu B := (1 - \nu) A + \nu B,$$

the *weighted operator arithmetic mean*, and

$$A \sharp_\nu B := A^{1/2} \left(A^{-1/2} B A^{-1/2}\right)^\nu A^{1/2},$$

the *weighted operator geometric mean*, where $\nu \in [0, 1]$. When $\nu = \frac{1}{2}$, we write $A\nabla B$ and $A\sharp B$ for brevity, respectively.

The definition $A\sharp_\nu B$ can be extended accordingly for any real number ν.

The following inequality is well as the operator *Young inequality* or operator *ν-weighted arithmetic–geometric mean inequality*:

$$A\sharp_\nu B \leq A\nabla_\nu B \quad \text{for all } \nu \in [0, 1]. \tag{58}$$

For recent results on operator Young inequality, see Refs. [1,15, 24,25,33,34].

Let $p \in (-\infty, 0) \cup (1, \infty)$ and $\Phi : J \subset (0, \infty) \to \mathbb{R}$ be twice differentiable on $\mathring{J}$ and such that for some γ, Γ, we have the condition (37). If we use the inequalities (28), (33) and (36), we have the positive invertible operators A and B that satisfy the condition

$$aA \le B \le bB$$

for some $0 < a < b$ then

$$\frac{\gamma}{p(p-1)} \left[\frac{1}{b-a} \left[a^p (bA - B) + b^p (B - aA) \right] - A \sharp_p B \right]$$

$$\le \frac{1}{b-a} \left[\Phi(a) (bA - B) + \Phi(b) (B - aA) \right] - \mathcal{P}_\Phi (B, A)$$

$$\le \frac{\Gamma}{p(p-1)} \left[\frac{1}{b-a} \left[a^p (bA - B) + b^p (B - aA) \right] - A \sharp_p B \right], \quad (59)$$

$$\frac{\gamma}{p(p-1)} \left[L_p^p(a,b) A - A \sharp_p B - p \left(A \sharp_{p-1} B \right) \left(\frac{a+b}{2} I - A^{-1} B \right) \right]$$

$$\le \left(\frac{1}{b-a} \int_a^b \Phi(t) \, dt \right) A - \mathcal{P}_\Phi (B, A) - \mathcal{P}_{\Phi'(\cdot)\left(\frac{a+b}{2} - \cdot\right)} (B, A)$$

$$\le \frac{\Gamma}{p(p-1)} \left[L_p^p(a,b) A - A \sharp_p B - p \left(A \sharp_{p-1} B \right) \right.$$

$$\left. \times \left(\frac{a+b}{2} I - A^{-1} B \right) \right] \qquad (60)$$

and

$$\frac{\gamma}{p(p-1)} \left[\frac{1}{2} \left(A \sharp_p B + \frac{1}{b-a} \left[b^p (bA - B) + a^p (B - aA) \right] \right) \right.$$

$$\left. - L_p^p(a,b) A \right]$$

$$\le \frac{1}{2} \left(\mathcal{P}_\Phi (B, A) + \frac{1}{b-a} \left[\Phi(b) (bA - B) + \Phi(a) (B - aA) \right] \right)$$

$$- \left(\frac{1}{b-a} \int_a^b \Phi(s) \, ds \right) A$$

$$\leq \frac{\Gamma}{p(p-1)} \left[\frac{1}{2} \left(A \sharp_p B + \frac{1}{b-a} \left[b^p \left(bA - B \right) + a^p \left(B - aA \right) \right] \right) \right.$$

$$\left. - L_p^p(a,b)A \right]. \tag{61}$$

Let $q \in (0,1)$ and $\Phi : J \subset (0,\infty) \to \mathbb{R}$ be twice differentiable on $\mathring{J}$ and such that for some γ, Γ, we have the condition (45). Then

$$\frac{\gamma}{q(1-q)} \left[A \sharp_q B - \frac{1}{b-a} \left[a^q \left(bA - B \right) + b^q \left(B - aA \right) \right] \right]$$

$$\leq \frac{1}{b-a} \left[\Phi \left(a \right) \left(bA - B \right) + \Phi \left(b \right) \left(B - aA \right) \right] - \mathcal{P}_\Phi \left(B, A \right)$$

$$\leq \frac{\Gamma}{q(1-q)} \left[A \sharp_q B - \frac{1}{b-a} \left[a^q \left(bA - B \right) + b^q \left(B - aA \right) \right] \right] \tag{62}$$

$$\frac{\gamma}{q(1-q)} \left[A \sharp_q B - q \left(A \sharp_{q-1} B \right) \left(\frac{a+b}{2} I - A^{-1} B \right) - L_q^q(a,b)A \right]$$

$$\leq \left(\frac{1}{b-a} \int_a^b \Phi \left(t \right) dt \right) A - \mathcal{P}_\Phi \left(B, A \right) - \mathcal{P}_{\Phi' (\cdot) \left(\frac{a+b}{2} - \cdot \right)} \left(B, A \right)$$

$$\leq \frac{\Gamma}{q(1-q)} \left[A \sharp_q B - q \left(A \sharp_{q-1} B \right) \left(\frac{a+b}{2} I - A^{-1} B \right) \right.$$

$$\left. - L_q^q(a,b)A \right] \tag{63}$$

and

$$\frac{\gamma}{q(1-q)} \left[L_q^q(a,b)A - \frac{1}{2} \left(A \sharp_q B + \frac{1}{b-a} \left[b^q \left(bA - B \right) \right. \right. \right.$$

$$\left. \left. \left. + a^q \left(B - aA \right) \right] \right) \right]$$

$$\leq \frac{1}{2} \left(\mathcal{P}_\Phi \left(B, A \right) + \frac{1}{b-a} \left[\Phi \left(b \right) \left(bA - B \right) + \Phi \left(a \right) \left(B - aA \right) \right] \right)$$

$$- \left(\frac{1}{b-a} \int_a^b \Phi \left(s \right) ds \right) A$$

$$\leq \frac{\Gamma}{q\,(1-q)} \left[L_q^q(a,b)A \right.$$

$$\left. -\frac{1}{2}\left(A\sharp_q B + \frac{1}{b-a}\left[b^q\,(bA - B) + a^q\,(B - aA) \right] \right) \right] \tag{64}$$

for positive invertible operators A and B that satisfy the condition (27).

5. Applications for Logarithm

Let $\Phi : J \subset (0,\infty) \to \mathbb{R}$ be twice differentiable on $\mathring{J}$ and such that for some δ, Δ, we have

$$\delta x^{-2} \leq \Phi''(x) \leq \Delta x^{-2} \quad \text{for any } x \in \mathring{J}. \tag{65}$$

We observe that the functions $\Phi - \delta\,(-\ln)$ and $\Delta\,(-\ln) - \Phi$ are convex functions on J. Since $\Psi := -\ln$ is also a convex function, it follows that $\Phi \in \mathcal{B}\,(J, m, M, \Psi)$ with $m = \delta$ and $M = \Delta$.

We define the weighted arithmetic mean and geometric mean as follows:

$$A_\nu\,(a,b) := (1-\nu)\,a + \nu b \quad \text{and} \quad G_\nu\,(a,b) := a^{1-\nu}b^\nu.$$

If we use the inequality (12), we have

$$\ln\left(\frac{A_\nu\,(a,b)}{G_\nu\,(a,b)} \right)^\delta \leq (1-\nu)\,\Phi\,(a) + \nu\Phi\,(b) - \Phi\,((1-\nu)\,a + \nu b)$$

$$\leq \ln\left(\frac{A_\nu\,(a,b)}{G_\nu\,(a,b)} \right)^\Delta \tag{66}$$

for any $a, b \in J$ and any $\nu \in [0,1]$.

From (18) and (19), we have

$$\ln\left(\frac{I\,(a,b)}{G\,(a,b)} \right)^\delta \leq \frac{\Phi\,(a) + \Phi\,(b)}{2} - \frac{1}{b-a}\int_a^b \Phi\,(t)\,dt$$

$$\leq \ln\left(\frac{I\,(a,b)}{G\,(a,b)} \right)^\Delta \tag{67}$$

and

$$\ln \left(\frac{A\,(a,b)}{I\,(a,b)} \right)^{\delta} \leq \frac{1}{b-a} \int_a^b \Phi\,(t)\,dt - \Phi\left(\frac{a+b}{2} \right)$$

$$\leq \ln \left(\frac{A\,(a,b)}{I\,(a,b)} \right)^M. \tag{68}$$

From the inequality (26), we have

$$\ln \left(\frac{I\,(a,b)}{\sqrt{A\,(a,b)\,G\,(a,b)}} \right)^{\delta}$$

$$\leq \frac{1}{2} \left(\Phi\left(\frac{a+b}{2} \right) + \frac{\Phi\,(b) + \Phi\,(a)}{2} \right) - \frac{1}{b-a} \int_a^b \Phi\,(s)\,ds$$

$$\leq \ln \left(\frac{I\,(a,b)}{\sqrt{A\,(a,b)\,G\,(a,b)}} \right)^{\Delta}. \tag{69}$$

For similar results see Ref. [5].

Fujii and Kamei [20,21] defined the *relative operator entropy* $S\,(A|B)$ for positive invertible operators A and B by

$$S\,(A|B) := A^{\frac{1}{2}} \left(\ln A^{-\frac{1}{2}} B A^{-\frac{1}{2}} \right) A^{\frac{1}{2}}, \tag{70}$$

which is a relative version of the operator entropy considered by Nakamura and Umegaki [32].

For some recent results on relative operator entropy, see Refs. [12, 13,26,27,29,30].

Consider the logarithmic function ln. Then the relative operator entropy can be interpreted as the perspective of ln, namely,

$$\mathcal{P}_{\ln}\,(B, A) = S\,(A|B).$$

If we use the inequalities (28), (33) and (36), we have the positive invertible operators A and B that satisfy the condition $aA \leq B \leq bB$, then we have

$$\delta \left[S\left(A|B\right) - \frac{1}{b-a} \left[\ln a \left(bA - B\right) + \ln b \left(B - aA\right)\right] \right]$$

$$\leq \frac{1}{b-a} \left[\Phi\left(a\right)\left(bA - B\right) + \Phi\left(b\right)\left(B - aA\right)\right] - \mathcal{P}_\Phi\left(B, A\right)$$

$$\leq \Delta \left[S\left(A|B\right) - \frac{1}{b-a} \left[\ln a \left(bA - B\right) + \ln b \left(B - aA\right)\right] \right], \quad (71)$$

$$\delta \left[S\left(A|B\right) + \frac{a+b}{2} AB^{-1}A - \ln I\left(a, b\right) A - A \right]$$

$$\leq \left(\frac{1}{b-a} \int_a^b \Phi\left(t\right) dt \right) A - \mathcal{P}_\Phi\left(B, A\right) - \mathcal{P}_{\Phi'\left(\cdot\right)\left(\frac{a+b}{2} - \cdot\right)}\left(B, A\right)$$

$$\leq \Delta \left[S\left(A|B\right) + \frac{a+b}{2} AB^{-1}A - \ln I\left(a, b\right) A - A \right] \quad (72)$$

and

$$\delta \Big[\ln I\left(a, b\right) A$$

$$- \frac{1}{2} \left(S\left(A|B\right) + \frac{1}{b-a} \left[\ln b \left(bA - B\right) + \ln a \left(B - aA\right)\right] \right) \Big]$$

$$\leq \frac{1}{2} \left(\mathcal{P}_\Phi\left(B, A\right) + \frac{1}{b-a} \left[\Phi\left(b\right)\left(bA - B\right) + \Phi\left(a\right)\left(B - aA\right)\right] \right)$$

$$- \left(\frac{1}{b-a} \int_a^b \Phi\left(s\right) ds \right) A$$

$$\leq \Delta \Big[\ln I\left(a, b\right) A - \frac{1}{2} \left(S\left(A|B\right) + \frac{1}{b-a} [\ln b \left(bA - B\right)\right.$$

$$\left. + \ln a \left(B - aA\right)] \right) \Big]. \quad (73)$$

Let $\Phi : J \subset (0, \infty) \to \mathbb{R}$ be twice differentiable on $\mathring{J}$ and such that for some θ, Θ, we have

$$\theta x^{-1} \leq \Phi''\left(x\right) \leq \Theta x^{-1} \quad \text{for any } x \in \mathring{J}. \quad (74)$$

We observe that the functions $\Phi - \theta \ell \ln$ and $\Theta \ell \ln - \Phi$ are convex functions on J, where ℓ is the identity function $\ell\left(t\right) = t$. Since

$\Psi := \ell \ln$ is also a convex function, it follows that $\Phi \in \mathcal{B}(J, m, M, \Psi)$ with $m = \theta$ and $M = \Theta$.

If we consider the entropy function $\eta(t) = -t \ln t$, then it is well known that for any positive invertible operators A, B, we have

$$S(A|B) = B^{1/2} \eta \left(B^{-1/2} A B^{-1/2} \right) B^{1/2}. \qquad (75)$$

The function $f(t) = t \ln t = -\eta(t)$, $t > 0$, is convex, then the perspective of this function is

$$\mathcal{P}_{(\cdot)\ln(\cdot)}(B, A) = -A^{1/2} \eta \left(A^{-1/2} B A^{-1/2} \right) A^{1/2} = -S(B|A),$$

where for the last equality we used (75) for A replacing B.

From the inequality (28), we have, for the convex function $\Psi(t) = t \ln t$, $t > 0$,

$$\theta \left[\frac{1}{b-a} \left[a \ln a \, (bA - B) + b \ln b \, (B - aA) \right] + S(B|A) \right]$$

$$\leq \frac{1}{b-a} \left[\Phi(a)(bA - B) + \Phi(b)(B - aA) \right] - \mathcal{P}_\Phi(B, A)$$

$$\leq \Theta \left[\frac{1}{b-a} \left[a \ln a \, (bA - B) + b \ln b \, (B - aA) \right] + S(B|A) \right], \qquad (76)$$

provided that $\Phi : J \subset (0, \infty) \to \mathbb{R}$ is twice differentiable on $\mathring{J}$ and such that the condition (74) holds while the operators A, B satisfy the condition (27) with $[a, b] \subset \mathring{J}$.

Similar inequalities can be stated by utilizing the results in (33) and (36), however the details are not presented here.

References

[1] H. Alzer, C. M. da Fonseca, and A. Kovačec (2015). Young-type inequalities and their matrix analogues. *Linear Multilinear Algebra*, **63**(3), 622–635. doi: 10.1080/03081087.2014.891588.

[2] S. S. Dragomir (2001). On a reverse of Jessen's inequality for isotonic linear functionals. *J. Inequal. Pure Appl. Math.*, **2**(3), 36.

[3] S. S. Dragomir (2001). Some inequalities for (m, M)-convex mappings and applications for the Csiszár Φ-divergence in information theory. *Math. J. Ibaraki Univ.*, **33**, 35–50. Preprint RGMIA Monographs, http://rgmia.org/papers/Csiszar/ImMCMACFDIT.pdf.

[4] S. S. Dragomir (2006). Bounds for the normalized Jensen functional. *Bull. Austral. Math. Soc.*, **74**(3), 417–478.

[5] S. S. Dragomir (2016). A survey on Jessen's type inequalities for positive functionals. In *Nonlinear Analysis*, P. M. Pardalos *et al.* (eds.), Springer Optimization and Its Applications, Vol. 68, In Honor of Themistocles M. Rassias on the Occasion of his 60th Birthday, doi: 10.1007/978-1-4614-3498-6_12. (Springer, New York, NY), pp. 177–232

[6] S. S. Dragomir (2015). A note on Young's inequality. *RGMIA Res. Rep. Coll.*, **18**, 126. http://rgmia.org/papers/v18/v18a126.pdf.

[7] S. S. Dragomir (2015). A note on new refinements and reverses of Young's inequality. *RGMIA Res. Rep. Coll.*, **18**, 131. http://rgmia.org/papers/v18/v18a131.pdf.

[8] S. S. Dragomir (2016). Additive inequalities for weighted harmonic and arithmetic operator means. *RGMIA Res. Rep. Coll.*, **19**, 6. http://rgmia.org/papers/v19/v19a06.pdf.

[9] S. S. Dragomir (2012). *Operator Inequalities of the Jensen, Čebyšev and Grüss Type.* Springer Briefs in Mathematics. (Springer, New York), xii+121 pp.

[10] S. S. Dragomir (2015). Some new reverses of Young's operator inequality, *RGMIA Res. Rep. Coll.*, **18**, 130. http://rgmia.org/papers/v18/v18a130.pdf.

[11] S. S. Dragomir (2015). On new refinements and reverses of Young's operator inequality. *RGMIA Res. Rep. Coll.*, **18**, 135. http://rgmia.org/papers/v18/v18a135.pdf.

[12] S. S. Dragomir (2015). Some inequalities for relative operator entropy. *RGMIA Res. Rep. Coll.*, **18**, 145. http://rgmia.org/papers/v18/v18a145.pdf.

[13] S. S. Dragomir (2015). Further inequalities for relative operator entropy. *RGMIA Res. Rep. Coll.*, **18**, 160. http://rgmia.org/papers/v18/v18a160.pdf.

[14] S. S. Dragomir and N. M. Ionescu (1990). On some inequalities for convex-dominated functions. *Mathematica — Rev. Numér. Theor. Approx.* **19**(1), 21–27.

[15] S. Furuichi (2012). Refined Young inequalities with Specht's ratio. *J. Egyptian Math. Soc.*, **20**, 46–49.

[16] S. Furuichi (2011). On refined Young inequalities and reverse inequalities. *J. Math. Inequal.*, **5**, 21–31.

[17] A. Ebadian, I. Nikoufar, and M. E. Gordji (2011). Perspectives of matrix convex functions. *Proc. Natl. Acad. Sci. USA*, **108**(18), 7313–7314.

[18] E. G. Effros (2009). A matrix convexity approach to some celebrated quantum inequalities. *Proc. Natl. Acad. Sci. USA*, **106**, 1006–1008.

[19] E. G. Effros and F. Hansen (2014). Noncomutative perspectives, *Ann. Funct. Anal.* **5**(2), 74–79.

[20] J. I. Fujii and E. Kamei (1989). Uhlmann's interpolational method for operator means. *Math. Jpn.*, **34**(4), 541–547.

[21] J. I. Fujii and E. Kamei (1989). Relative operator entropy in noncommutative information theory. *Math. Jpn.*, **34**(3), 341–348.

[22] S. Furuichi and N. Minculete (2011). Alternative reverse inequalities for Young's inequality. *J. Math Inequal.*, **5**(4), 595–600.

[23] F. Kittaneh and Y. Manasrah (2018). Improved Young and Heinz inequalities for matrix. *J. Math. Anal. Appl.*, **361**, 262–269.

[24] F. Kittaneh and Y. Manasrah (2011). Reverse Young and Heinz inequalities for matrices. *Linear Multilinear Algebra*, **59**, 1031–1037.

[25] W. Liao, J. Wu, and J. Zhao (2015). New versions of reverse Young and Heinz mean inequalities with the Kantorovich constant. *Taiwanese J. Math.*, **19**(2), pp. 467–479.

[26] I. H. Kim (2012). Operator extension of strong subadditivity of entropy, *J. Math. Phys.*, **53**, 122204.

[27] P. Kluza and M. Niezgoda (2014). Inequalities for relative operator entropies, *Electron. J. Linear Algebran*, **27**, 1066.

[28] F. Kubo and T. Ando (1980). Means of positive operators. *Math. Ann.*, **264**, 205–224.

[29] M. S. Moslehian, F. Mirzapour, and A. Morassaei (2013). Operator entropy inequalities. *Colloq. Math.*, **130**, 159–168.

[30] I. Nikoufar (2014). On operator inequalities of some relative operator entropies. *Adv. Math.*, **259**, 376–383.

[31] A. Ostrowski (1983). Über die Absolutabweichung einer differentiierbaren Funktion von ihrem Integralmittelwert. *Comment. Math. Helv.*, **10**, 226–227.

[32] M. Nakamura and H. Umegaki (1961). A note on the entropy for operator algebras. *Proc. Japan Acad.*, **37**, 149–154.

[33] M. Tominaga (2002). Specht's ratio in the Young inequality. *Sci. Math. Jpn.*, **55**, 583–588.

[34] G. Zuo, G. Shi, and M. Fujii (2011). Refined Young inequality with Kantorovich constant, *J. Math. Inequal.*, **5**, 551–556.

ientific Publishing Company
.../ 10.1142/9789811271922_0005

Chapter 5

Product Subset Problem: Applications to Number Theory and Cryptography*

Konstantinos A. Draziotis[†], Vasileios Martidis[‡] and Stratos Tiganourias[§]

Department of Computer Science, Aristotle University of Thessaloniki, 54124 Thessaloniki, Greece

[†] *drazioti@csd.auth.gr*
[‡] *vamartid@yandex.com*
[§] *etiganou97@gmail.com*

We consider the applications of subset product problem (SPP) in number theory and cryptography. We obtain a probabilistic algorithm that solves SPP and we analyze it with respect to time–space complexity and success probability. Furthermore, we provide an application to the problem of finding Carmichael numbers with many factors and an attack to Naccache–Stern knapsack cryptosystem, where we update previous results.

1. Introduction

In this chapter, we study the modular version of subset product problem (MSPP). We shall provide an algorithm for solving MSPP based

---

*All the authors contributed equally to this research.

91

on birthday paradox. What is more, we analyze the algorithm with respect to success probability and time–space complexity. Our applications concern the following two problems: find Carmichael numbers with many prime factors and as a second application, provide an attack to Naccache–Stern knapsack (NSK) public key cryptosystem. We begin with the definition of product subset problem.

Definition 1 (Subset Product Problem (SPP)). Given a list of integers L and an integer c, find a subset of L whose product is c.

This problem is (strong) NP-complete using a transformation from the exact cover by 3-sets (X3C) problem (See Refs. [9, p. 224] and [33]). Also, in Ref. [8], Theorem 3.2, the authors proved that SPP is at least as hard as the *clique* problem (with respect to fixed-parameter tractability). In this chapter, we consider the following variant.

Definition 2 (Modular Subset Product Problem: MSPP$_\Lambda$). Given a positive integer Λ, an integer $c \in \mathbf{Z}_\Lambda^*$ and a vector $(u_0, u_1, \ldots, u_n) \in (\mathbf{Z}_\Lambda^*)^{n+1}$, find a binary vector $m = (m_0, m_1, \ldots, m_n)$ such that

$$c \equiv \prod_{i=0}^{n} u_i^{m_i} \bmod \Lambda. \tag{1}$$

The MSPP$_\Lambda(\mathcal{P}, c)$ problem can be defined equivalently as follows: Given a finite set $\mathcal{P} \subset \mathbf{Z}_\Lambda^*$ and a number $c \in \mathbf{Z}_\Lambda^*$, find a subset $\mathcal{B}$ of $\mathcal{P}$, such that

$$\prod_{x \in \mathcal{B}} x \equiv c \bmod \Lambda.$$

What is more, we can define MSPP for a general abelian finite group G as follows. We write G multiplicative.

Definition 3 (Modular Subset Product Problem for G: MSPP$_G(\mathcal{P}, c)$). Given an element $c \in G$ and a vector $(u_0, u_1, \ldots, u_n) \in G^{n+1}$, find a binary vector $m = (m_0, \ldots, m_n)$ such that

$$c = \prod_{i=0}^{n} u_i^{m_i}. \tag{2}$$

However, in this work, we consider the case $G = \mathbf{Z}_Q^*$, where Q is highly composite number (the case of Carmichael numbers) or prime (the case of NSK cryptosystem).

1.1. *Our contribution*

First, we provide an algorithm for solving product modular subset problem based on birthday paradox. Our approach is not new, for instance, see Ref. [24, Section 2.3]. Here, we use a variant of Ref. [4, Section 3]. We implement a parallel version of this algorithm and provide an analysis concerning complexity issues and its success probability. This algorithm is applied to the cryptanalysis of NSK, thus we update the tables provided in Ref. [4]. Furthermore, except the cryptanalysis of NSK cryptosystem, we apply our algorithm to the search of Carmichael numbers. In order to achieve this, we use an old method of Erdős. For instance, we generate a Carmichael number with 19589 prime factors.[1] Finally, we provide an abstract version of the algorithm (see Ref. [4, Section 3]) to the general product subset problem. In advance, we analyze the algorithm as far as the selection of the parameters (this is provided in Proposition 1).

Roadmap: This chapter is organized as follows. In Sections 2 and 3, we introduce the birthday paradox attack to MSPP. Section 4 is dedicated to applications. In Section 4.1, we use MSPP in order to attack NSK cryptosystem and in Section 4.3, we provide some experimental results. Section 4.4 is dedicated to the problem of finding Carmichael numbers with many prime factors. We provide the necessary bibliography and known results. Finally, the last section contains some concluding remarks.

2. Birthday Attack to Modular Subset Product Problem

We call density of $\mathrm{MSPP}_G(\mathcal{P}, c)$ the positive real number

$$d = \frac{|\mathcal{P}|}{\log_2 |G|}.$$

---

[1]See http://tiny.cc/tm6miz.

If $G = \mathbf{Z}_\Lambda^*$, then

$$d = \frac{|\mathcal{P}|}{\log_2 |\mathbf{Z}_\Lambda^*|} = \frac{|\mathcal{P}|}{\log_2 \phi(\Lambda)},$$

where ϕ is the Euler totient function. In a $\mathrm{MSPP}_G(\mathcal{P}, c)$ having a large density, we expect to have many solutions.

A straightforward attack uses birthday paradox paradigm to $\mathrm{MSPP}_\Lambda(\mathcal{P}, c)$. Rewriting equivalence (1) as

$$\prod_{i=0}^{\alpha} u_i^{m_i} \equiv c \prod_{i=\alpha+1}^{n} u_i^{-m_i} \bmod \Lambda,$$

for some integer $\alpha \approx n/2$, we construct two subsets of $\mathbb{Z}_\Lambda$, say U_1 and U_2. The first contains elements of the form $\prod_{i=0}^{\alpha} u_i^{m_i} \bmod \Lambda$ and the second $c \prod_{i=\alpha+1}^{n} u_i^{-m_i} \bmod \Lambda$ for all possible (binary) values of $\{m_i\}_i$. So, the problem reduces to finding a common element of sets U_1 and U_2. In the following, we provide the pseudocode of the previous algorithm.

Algorithm 1. Birthday attack to $\mathrm{MSPP}_\Lambda(\mathcal{P}, c)$

INPUT : $\mathcal{P} = \{u_i\}_i \subset \mathbf{Z}_\Lambda^*$ ($|\mathcal{P}| = n + 1$), $c \in \mathbf{Z}_\Lambda^*$ (assume that $\gcd(u_i, \Lambda) = 1$)
OUTPUT: A set $\mathcal{B} \subset \mathcal{P}$ such that, $\prod_{x \in \mathcal{B}} x \equiv c \bmod \Lambda$ or Fail

1: $I_1 \leftarrow \{0, 1, \ldots, \lceil n/2 \rceil\}, I_2 \leftarrow \{\lceil n/2 \rceil + 1, \ldots, n\}$
2: $U_1 \leftarrow \left\{ \prod_{i \in I_1} u_i^{\varepsilon_i} \bmod \Lambda : \text{for all } \varepsilon_i \in \{0, 1\} \right\}$
3: $U_2 \leftarrow \left\{ c \prod_{i \in I_2} u_i^{-\varepsilon_i} \bmod \Lambda : \text{for all } \varepsilon_i \in \{0, 1\} \right\}$
4: **If** $U_1 \cap U_2 \neq \emptyset$
5:     Let y be an element of $U_1 \cap U_2$
6:     **return** $u_i : \prod u_i \equiv c \bmod \Lambda$.
7: **else return** Fail: There is not any solution

This algorithm is deterministic. Indeed, if there is a solution to $\mathrm{MSPP}_\Lambda(\mathcal{P}, c)$, then the algorithm will find it. To construct the solution in step 6, we use the following equation:

$$\prod_{i \in I_1} u_i^{\varepsilon_i} = c \prod_{j \in I_2} u_j^{-\varepsilon_j} \quad (\text{in } \mathbf{Z}_\Lambda). \tag{3}$$

It turns out that $y = \prod_{i=0}^{n} u_i^{\varepsilon_i} \bmod \Lambda$. For storage, we need $2^{n/2+1}$ elements of $\mathbf{Z}_\Lambda^*$. In line 4, we compute a common element from U_1 and U_2. To do this, we first sort the elements of U_1, U_2 and then we apply binary search. Overall, we need $O(2^{n/2} \log_2 n)$ arithmetic operations in the multiplicative group $\mathbf{Z}_\Lambda^*$. The drawback of this algorithm is the large space complexity. We can improve the previous algorithm as far as the space complexity.

3. An Improvement of Algorithm 1

We provide the following definitions.

Definition 4. Let $c \in \mathbf{Z}_\Lambda^*$. We define

$$\text{sol}(c; \mathcal{P}, \Lambda) = \left\{ I \subset \{0, 1, \dots, n\} : c \equiv \prod_{i \in I, \; u_i \in \mathcal{P}} u_i \bmod \Lambda \right\}.$$

Let the map,

$$\chi : \text{sol}(c; \mathcal{P}, \Lambda) \to \{0, 1\}^{n+1},$$

such that $\chi(I) = (\varepsilon_0, \varepsilon_1, \dots, \varepsilon_n)$, where $\varepsilon_i = 1$ if $i \in I$ else $\varepsilon_i = 0$.

Definition 5. We define

$$\text{Sol}(c; \mathcal{P}, \Lambda) = \{(u_0^{\varepsilon_0}, u_1^{\varepsilon_1}, \dots, u_n^{\varepsilon_n}) : \chi(I) = (\varepsilon_0, \varepsilon_1, \dots, \varepsilon_n)$$
$$\text{for all } I \in \text{sol}(c; \mathcal{P}, \Lambda)\}.$$

Definition 6. To each element I of $\text{sol}(c; \mathcal{P}, \Lambda)$ (assuming there exists one), we map the natural number $H_I(c) = |I|$ (the cardinality of I). We call this number local Hamming weight of c at I. We call Hamming weight $H(c)$ of c, the minimum of all these numbers:

$$H(c) = \min\{H_I(c) : I \in \text{Sol}(c; \mathcal{P}, \Lambda)\}.$$

Remark that the local Hamming weight of c at I is in fact the Hamming weight of the binary vector $\chi(I)$, i.e. the sum of all the entries of $\chi(I)$. Thus,

$$H_I(c) = \sum_{i=0}^{n} \varepsilon_i, \quad \text{where } \chi(I) = (\varepsilon_0, \varepsilon_1, \dots, \varepsilon_n).$$

Example 1. Let $\mathcal{P} = \{2, 3, \ldots, 10^7\}$, $\Lambda = 10000019$, $c = 190238$. There is an element Σ of $\text{Sol}(c; \mathcal{P}, \Lambda)$ (with $|\Sigma| = 11$)

$$\Sigma = (9851537, 303860, 4680021, 9647209, 2006838, 9984877,$$
$$2512434, 2126904, 1942182, 8985302, 2193757).$$

Note that we have trimmed all the ones:

$$\Sigma' = (1, \ldots, 1, 303860, 1, \ldots, 1942182, 1, \ldots, 9984877, 1, \ldots, 1).$$

Straightforward calculations provide

$$\prod_{x \in \Sigma} x = \prod_{x \in \Sigma'} x \equiv 190238 = c \pmod{\Lambda}.$$

Further, there is an element in $\text{sol}(c; \mathcal{P}, \Lambda)$ say I, such that

$$\chi(I) = \Sigma'.$$

So, in this case, the local Hamming weight of c at I is 11. However, the computation of the Hamming weight of c is more difficult. Note that $H(c) \geq 2$, except if some u_i is equal to c. We ignore this trivial case. For now, we can only say that $H(c) \leq 11$, i.e. every local Hamming weight is an upper bound for $H(c)$.

Now, let $I \in \text{sol}(c; \mathcal{P}, \Lambda)$. We consider two positive integers, say h_1, h_2, such that $h_1 + h_2 = H_I(c)$, and two disjoint subsets I_1, I_2 of $\{0, 1, \ldots, n\}$ with $|I_1| = |I_2| = b$, for some positive integer $b \leq n/2$. Finally, we consider the sets,

$$U_{h_1}(I_1; \mathcal{P}, \Lambda) = \left\{ \prod_{i \in I_1} u_i^{\varepsilon_i} \pmod{\Lambda} : \sum_{i \in I_1} \varepsilon_i = h_1 \right\},$$

$$U_{h_2}(I_2, c; \mathcal{P}, \Lambda) = \left\{ c \prod_{i \in I_2} u_i^{-\varepsilon_i} \pmod{\Lambda} : \sum_{i \in I_2} \varepsilon_i = h_2 \right\}.$$

We usually write them as $U_{h_1}(I_1)$ and $U_{h_2}(I_2, c)$ since $\mathcal{P}$ and Λ are known. We have

$$|U_{h_1}(I_1)| = \binom{|I_1|}{h_1}, \quad |U_{h_2}(I_2, c)| = \binom{|I_2|}{h_2}.$$

Remark 1. The set U_1 of Algorithm 1 is written as

$$U_1 = \bigcup_{h_1=1}^{\lceil n/2 \rceil + 1} U_{h_1}(\{0, 1, \ldots, \lceil n/2 \rceil\}; \mathcal{P}, \Lambda).$$

That is, U_1 is the union of the sets $\{U_h(I; \mathcal{P}, \Lambda)\}_{1 \le h \le \lceil n/2 \rceil + 1}$ for $I = \{0, 1, \ldots, \lceil n/2 \rceil\}$. Similarly, U_2 is written as

$$U_2 = \bigcup_{h_2=0}^{n - \lceil \frac{n}{2} \rceil} U_{h_2}(\{\lceil n/2 \rceil + 1, \ldots, n\}, c; \mathcal{P}, \Lambda).$$

Instead of using U_1 and U_2, we use specific subsets of them. The choice of subsets creates a probability distribution at the output. Thus, this choice must be studied as far as the success probability of the algorithm. For the following algorithm, we assume that we know a local Hamming weight of the target number c.

Algorithm 2. BA_MSPP$_\Lambda(\mathcal{P}, c; b, \ell, Q, \text{iter})^2$: Memory efficient attack to MSPP$_\Lambda(\mathcal{P}, c)$
INPUT:

i. A set $\mathcal{P} = \{u_i\}_i \subset \mathbf{Z}_\Lambda^*$ with $|\mathcal{P}| = n + 1$ (assume that $\gcd(u_i, \Lambda) = 1$)
ii. a number $c \in \mathbf{Z}_\Lambda^*$
iii. a local Hamming weight of c, say ℓ
iv. a positive number b: $\ell \le b \le n/2$
v. a compression function $\mathbb{H}$
vi. a positive integer *iter*

OUTPUT: a set $\mathcal{B} \subset \mathcal{P}$, such that $\prod_{x \in \mathcal{B}} x \equiv c \bmod \Lambda$ or Fail

1: $(h_1, h_2) \leftarrow (\lfloor \ell/2 \rfloor, \lceil \ell/2 \rceil)$
2: **For** i in $1, \ldots, \text{iter}$
3:     $(I_1, I_2) \xleftarrow{\$} \{0, \ldots, n\} \times \{0, \ldots, n\}$ # with I_1, I_2 disjoint and
      $|I_1| = |I_2| = b$
4:     $U_{h_1}^*(I_1) \leftarrow \left\{ \mathbb{H}\left(\prod_{i \in I_1} u_i^{\varepsilon_i} \pmod{\Lambda} \right) : \sum_{i \in I_1} \varepsilon_i = h_1, \ \varepsilon_i \in \{0, 1\} \right\}$

---

[2]BA: Birthday Attack.

5:     **For** each $(\varepsilon_i)_i$ such that : $\sum_{i \in I_2} \varepsilon_i = h_2$
6:         **If** $\mathbb{H}\left(c \prod_{i \in I_2} u_i^{-\varepsilon_i} \pmod{\Lambda}\right) \in U_{h_1}^*(I_1)$
7:             Let y be an element of $U_{h_1}^*(I_1) \cap U_{h_2}^*(I_2, c)$
8:                 **return** $\{u_i\}_i$ such that $\prod u_i \equiv c \bmod \Lambda$ **and terminate**
9: **return** Fail # if for all the iterations the algorithm fails to find
a solution

This algorithm is a memory-efficient version of Algorithm 1, since
we consider subsets of U_1, U_2. However, this algorithm may fail, even
when $\mathrm{MSPP}_\Lambda(\mathcal{P}, c)$ has a solution. For instance, we assume that
$\mathrm{sol}(c; \mathcal{P}, \Lambda) \neq \emptyset$. If we pick I_1, I_2 and it happens that the union $I_1 \cup$
$I_2 \not\subseteq \mathrm{sol}(c; \mathcal{P}, \Lambda)$, then the algorithm will fail, i.e. the initial problem
may have a solution and the algorithm fails to find it. This may occur
when $b < n/2$, that is $I_1 \cup I_2 \subset \{0, 1, \ldots, n\}$. If $I_1 \cup I_2 = \{0, 1, \ldots, n\}$,
then the algorithm remains probabilistic, since we consider a specific
choice of (h_1, h_2) and not all the possible (h_1, h_2), with $h_1 + h_2 = \ell$.
If we consider all (h_1, h_2) such that $h_1 + h_2 = n$, then the algorithm
turns out to be deterministic.
We analyze the algorithm line by line.

Line 3: This can be readily implemented in case of $2b < \sqrt{n}$. Indeed,
we can use rejection sampling in the set $\{0, \ldots, n\}$ and construct a
list of length $2b$. Then, I_1 is the set consisting of the first b elements
and I_2 consists of the rest elements. If $2b \geq \sqrt{n}$, then we have to
sample from the set $\{0, 1, \ldots, n\}$, so the memory increases since we
have to store the set $\{0, 1, \ldots, n\}$. For instance, when we apply the
algorithm in order to find Carmichael numbers, we have $2b \ll \sqrt{n}$.
In the case of NSK cryptosystem, we usually have $2b > \sqrt{n}$.

Line 4: The most intensive part (both for memory and time complexity) is the construction of the set $U_{h_1}^*$. Here, we can parallelize
our algorithm to decrease time complexity. To reduce the space complexity, we use the parameter $b \leq n/2$ and the compression function
$\mathbb{H}$. For instance, as a compression function, we consider Q-strings of
the output of a hash function. We suppose that the output of the
hash function is given as a hex string. In Section 3.4, we provide a
strategy to choose Q.

Lines 5–6: In line 4, we store $U_{h_1}^*(I_1)$; in this line, we compute *on
the fly* the elements of the second set $U_{h_2}^*(I_2, c)$ and check if any is in

$U^*_{h_1}(I_1)$. So, we do not need to store $U^*_{h_2}$. A suitable data structure for the search is a hashtable, which we also used in our implementation. Hashtables have the advantage of the fast insert, delete and search operations, since these operations have $O(1)$ time complexity on the average.

Lines 7–8: Having the element found by the previous step (line 6), say y, we construct the u_i's such that their product is y. We return Fail if the intersection is empty for all the iterations.

Remark 2. If we do not consider any hash function and iter $= 1$, then we write $\text{BA\_MSPP}_\Lambda(\mathcal{P}, c; b, \ell)$.

3.1. *Space complexity*

We shall compute an upper bound for the memory $\mathbb{M}$. We assume that there is not any collision in the construction of $U^*_{h_1}(I_1)$ and $U^*_{h_2}(I_2, c)$, or in other words, we choose Q to minimize the probability to have a collision, i.e. we choose $2^{4Q} \gg \max\{|U_1|, |U_2|\}$ and we assume that $\mathbb{H}$ behaves random enough. In practice (or at least in our examples), we always meet this constraint. So, we get

$$|U_{h_1}(I_1)| = |U^*_{h_1}(I_1)| = \binom{|I_1|}{h_1} \quad \text{and}$$

$$|U_{h_2}(I_2, c)| = |U^*_{h_2}(I_2, c)| = \binom{|I_2|}{h_2},$$

where $(h_1, h_2) = (\lfloor \ell/2 \rfloor, \lceil \ell/2 \rceil)$. By choosing $|I_1| = |I_2| = b$, we get

$$|U^*_{h_1}(I_1)| = \binom{b}{h_1} \quad \text{and} \quad |U^*_{h_2}(I_2, c)| = \binom{b}{h_2}.$$

We set

$$S_b = \binom{b}{h_1} = B_b(h_1).$$

In our algorithm, we need to store the S_b hashes of the set $U^*_{h_1}(I_1)$. We need $4Q$-bits for keeping Q-hex digits in the memory. So, overall, we store $4QS_b$ bits. Furthermore, we store the binary exponents $(\varepsilon_i)_i$ that are necessary for the computation of the products, $\prod_{i \in I_1} u_i^{\varepsilon_i}$.

Table 1. For $b = 50$, $Q = 12$ and $\mathcal{P} = \{2, 3, \ldots, 10^7\}$.

ℓ	9	11	13	15	17	19	21
M (GB)	0.029	0.05	0.2	1.16	6.15	28.61	117.2

This is needed, since we must reconstruct c as a product of $(u_i)_i$. These are S_b, thus we need $b \times S_b$-bits. Since we also need to store the set $\mathcal{P}$ of length $n + 1$, we conclude that

$$\mathbb{M} < (4Q + b)S_b + (n + 1)B \quad \text{(bits)}, \qquad (4)$$

where $B = \max\{\log_2(x) : x \in \mathcal{P}\}$. Remark that $\mathbb{M}$ does not depend on the modulus Λ.

If we can describe in an efficient way, the set $\mathcal{P}$ we do not need to store it. Say that $\mathcal{P} = \{2, 3, \ldots, n\}$. Then, there is no need to store it in the memory, since the sequence $f(x) = x + 1$ describes efficiently the set $\mathcal{P}$. Also, in other situations, B can be stored using $O(|\mathcal{P}| \log_2(|\mathcal{P}|))$ bits. We call such sets *nice* and they can save us enough memory. In fact, for nice sets, the inequality (4) changes to

$$\mathbb{M} < (4Q + b)S_b + O(n \log_2 n) \quad \text{(bits)}, \qquad (5)$$

where B has disappeared. B in some problems can be a very large number. In fact when we apply this algorithm to the problem of finding Carmichael numbers, we shall see that the set $\mathcal{P}$ is *nice*.

Finally, if U_{h_1} is very large, we can make chunks of it before we store it in the memory. Note that this cannot be done if we directly compute the intersection of $U_{h_1} \cap U_{h_2}$ as in Ref. [4]. This simple trick considerably improves the algorithms in Ref. [4].

3.2. *Time complexity*

Time complexity is dominated by the construction of the sets U_{h_1} and U_{h_2} and the calculation of their intersection. We work with U_{h_1} instead of $U_{h_1}^*$ since all the multiplications are between the elements of U_{h_1}, only in the searching phase we move to $U_{h_1}^*$. Let M_Λ be the bit complexity of the multiplication of two integers $\bmod \Lambda$. So,

$$M_\Lambda = O((\log_2 \Lambda)^{1+\varepsilon})$$

for some $0 < \varepsilon \leq 1$ (for instance, Karatsuba suggests $\varepsilon = \log_2 3 - 1$ [16]). In fact, recently, it was proved

$$M_\Lambda = O((\log_2 \Lambda)^{1+\varepsilon}),$$

see [13]. To construct the sets U_{h_1} and U_{h_2} (ignoring the cost for the inversion mod Λ), we need $M_\Lambda h_1 B_b(h_1) = M_\Lambda h_1 S_b$ bit operations for the set U_{h_1} and $M_\Lambda h_2 B_b(h_2 + 1) \approx M_\Lambda h_2 S_b$ bit operations for U_{h_2}. So, overall

$$\mathbb{T}_1 = M_\Lambda(h_1 S_b + h_2 S_b) = M_\Lambda S_b(h_1 + h_2) \text{ bits.}$$

Considering the time complexity for finding a collision in the two sets by using a hashtable, we get

$$\mathbb{T} = \mathbb{T}_1 + O(1)S_b \text{ bits on average}$$

and

$$\mathbb{T} = \mathbb{T}_1 + O(1)S_b^2 \text{ in the worst case.}$$

We used that $h_1 \approx h_2$ (they differ at most by 1). In case we have T threads, we get about $\mathbb{T}/T$ (bit operations) instead of $\mathbb{T}$.

3.3. *Success probability*

In the following lemma, we compute the probability to get a common element in $U_{h_1}(I_1)$ and $U_{h_2}(I_2, c)$ for $(I_1, I_2) \xleftarrow{\$} \{0, \ldots, n\} \times \{0, \ldots, n\}$, I_1, I_2 disjoints, each one with b elements. We assume that $0 \le y \le x$, then with $B_x(y)$, we denote the binomial coefficient $\binom{x}{y}$.

Lemma 1. *Let h_1, h_2 be positive integers and $\ell = h_1 + h_2$. The probability to get $U_{h_1}(I_1) \cap U_{h_2}(I_2, c) \ne \emptyset$ is*

$$\mathbb{P} = \frac{B_b(h_1)B_b(h_2)}{B_{n+1}(\ell)}.$$

For the proof, see Ref. [4, Section 3].

We can easily provide another and simpler proof in the case $2b = n + 1$ (this occurs very often when attacking Naccache–Stern cryptosystem). Then,

$$\mathbb{P} = \text{hyper}(x; 2b, b, \ell),$$

where hyper is the hypergeometric distribution,

$$\text{hyper}(x; N, b, \ell) = \Pr(X = x) = \frac{\binom{b}{x}\binom{N-b}{\ell-x}}{\binom{N}{\ell}},$$

where

- $N = n + 1$ is the population size,
- ℓ is the number of draws,
- b is the number of successes in the population,
- x is the number of observed successes.

In our case, we set $N = 2b = n + 1$, $\ell = h_1 + h_2$, and $x = h_1$. Then,

$$\text{hyper}(x = h_1; n + 1, b, \ell) = \Pr(X = h_1) = \frac{\binom{b}{h_1}\binom{b}{\ell - h_1}}{\binom{n+1}{\ell}} = \mathbb{P}.$$

The expected value is $\frac{b\ell}{n+1}$ and since $b = (n + 1)/2$, we get $EX = \frac{\ell}{2}$. Since the random variable X counts the successes, we expect on average to have $\ell/2$ successes after considering enough instances (i.e. choices of I_1, I_2). The maximum value of ℓ is $n/2$. Thus, in this case, the expected value is maximized, hence we expect our algorithm to find faster a solution from another one that uses smaller value for ℓ.

Also, we need about

$$\frac{1}{\mathbb{P}} = \frac{\binom{n+1}{\ell}}{\binom{b}{h_1}\binom{b}{h_2}} \text{ iterations on average to find a solution.}$$

One last remark is that the contribution of b is bigger than the contribution of Hamming weight in the probability $\mathbb{P}$ (see Figure 1).

3.3.1.   *The best choice of h_1 and h_2*

The choice of h_1, h_2 (in line 1 of Algorithm 2) is $h_1 = \lfloor \ell/2 \rfloor$, $h_2 = \lceil \ell/2 \rceil$. This can be explained easily, since these values maximize the probability $\mathbb{P}$ of Lemma 1. We set

$$J_\ell = \{(x, y) \in \mathbf{Z}^2 : x + y = \ell, 0 \leq x \leq y\}.$$

Observe that $(\lfloor \ell/2 \rfloor, \lceil \ell/2 \rceil) \in J_\ell$.

Proposition 1. *Let b and n be fixed positive integers such that $\ell = x + y \leq b \leq n/2$. Then the finite sequence $\mathbb{P} : J_\ell \to \mathbb{Q}$, defined by*

$$\mathbb{P}(x, y) = \frac{B_b(x)B_b(y)}{B_{n+1}(\ell)},$$

is maximized for $(x, y) = (\lfloor \ell/2 \rfloor, \lceil \ell/2 \rceil)$.

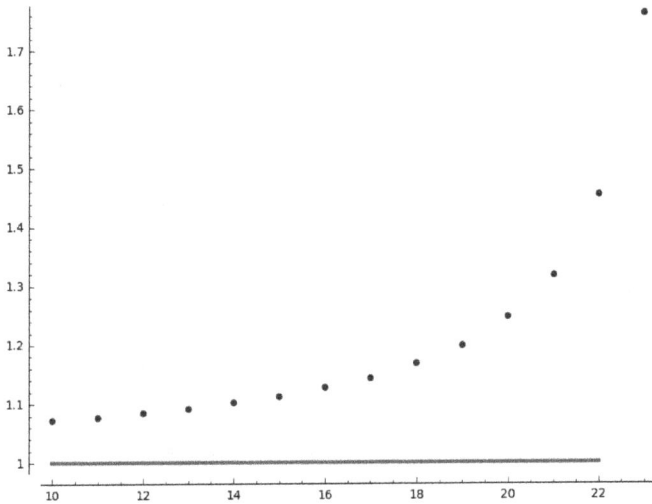

Figure 1. We fixed $n = 23000$ and $b = 35$. We consider pairs $(h, f(h))$, where $f(h) = (\mathbb{P}(b+1, h)/\mathbb{P}(b, h+1))$ and h the Hamming weight. We finally normalized the values by taking logarithms. So, for a given (b, h), if we want to increase the probability, it is better to increase b than h.

We need the following simple lemma.

Lemma 2.

$$\binom{b}{x}\binom{b}{\ell - x}\binom{2b}{b} = \binom{\ell}{x}\binom{2b - \ell}{b - x}\binom{2b}{\ell}.$$

Proof. It is straightforward by expressing the binomial coefficients in terms of factorials and rearranging them. $\qquad\square$

Proof of Proposition 1. From the previous lemma, we have

$$\frac{\binom{b}{x}\binom{b}{\ell - x}}{\binom{2b}{\ell}} = \frac{\binom{\ell}{x}\binom{2b - \ell}{b - x}}{\binom{2b}{b}}. \tag{6}$$

Since b, ℓ are fixed, the maximum of the right-hand side occurs at $x = \lfloor \ell/2 \rfloor$. Indeed, both sequences $\binom{b}{x}$ and $\binom{2b - \ell}{b - x}$ are positive and maximized at $x = \lfloor \ell/2 \rfloor$. Thus, the same holds for the product $\binom{\ell}{x}\binom{2b - \ell}{b - x}$, i.e. is maximized at $x = \lfloor \ell/2 \rfloor$. The same occurs to the left-hand side. Since the denominator of the left-hand side in (6) is fixed, the numerator $B_b(x)B_b(\ell - x) = \binom{b}{x}\binom{b}{\ell - x}$ is maximized at $x = \lfloor \ell/2 \rfloor$. Since n is also a fixed positive integer, the numerator of $\mathbb{P}$ is maximized at $x = \lfloor \ell/2 \rfloor$ and so $\mathbb{P}$ is maximized at the same point. Finally, $\ell - \lfloor \ell/2 \rfloor = \lceil \ell/2 \rceil = y$. The proposition follows. □

3.4.   How to choose Q?

If we are searching for r-same objects to one set (with cardinality n), when we pick the elements of the set from some largest set (with cardinality m), then we say that we have a r-multicollision. We have the following lemma.

Lemma 3. *If we have a set with m elements and we pick randomly (and independently) n elements from the set, then the expected number of r-multicollisions is approximately*

$$\frac{n^r}{r! m^{r-1}}. \tag{7}$$

Proof.   See Ref. [15, Section 6.2.1]. □

Say we use md5 hash function. If we use the parameter Q, we have to truncate the output of md5, which has 16-hex strings, to Q-hex strings ($Q < 16$), i.e. we only consider the first $\kappa = 4 \cdot Q$-bits of the output. Our strategy to choose Q uses formula (7). In practice, it is enough to avoid $r = 3$-multicollisions in the set $U^*_{h_1}(I_1) \cup U^*_{h_2}(I_2, c)$ of cardinality S_b, where

$$S_b = |U^*_{h_1}(I_1) \cup U^*_{h_2}(I_2, c)| = \binom{|I_1|}{h_1} + \binom{|I_1|}{h_2}.$$

We set the formula (7) equal to 1 and we solve with respect to m, which in our case is $m = 2^\kappa$. So, $\frac{S_b^3}{6} = 2^{2\kappa}$. Therefore, we get $\kappa \approx (3\log_2 S_b)/2$ (since in our examples, $\log_2(S_b) \gg \log_2 6$). For instance, if we have local Hamming weight ≤ 13, $b = n/2$, $|I_1| = |I_2| = b$, and $n = 232$, we get $\kappa \approx 52$. So, $Q \approx 13$. In fact, we used $Q = 12$ in our attack to NSK cryptosystem.

4. Applications

4.1. *Naccache–Stern knapsack cryptosystem*

In this section, we consider an application to cryptography. We shall provide an attack to a public key cryptosystem. NSK cryptosystem is a public key cryptosystem [24] based on the discrete logarithm problem (DLP), which is a difficult number theory problem. Furthermore, it is based on another combinatorial problem, the MSPP. Our attack applies to the latter problem. NSK cryptosystem is defined by the following three algorithms:

(i) Key Generation: Let p be a large safe prime number (that is $(p-1)/2$ is a prime number). Let n denotes the largest positive integer such that

$$p > \prod_{i=0}^{n} p_i, \tag{8}$$

where p_i is the $(i+1)$th prime. The message space of the system is $\mathcal{M} = \{0,1\}^{n+1}$, this is the set of the binary strings of $(n+1)$-bits. For instance, if p has 2048 bits, then $n = 232$ and if p has 1024 bits, then $n = 130$.

We randomly pick a positive integer $s < p-1$, such that $\gcd(s, p-1) = 1$. This last property guarantees that there exists the (unique) sth root $\bmod\, p$ of an element in $\mathbf{Z}_p^*$. Set

$$u_i = \sqrt[s]{p_i} \bmod p \in \mathbf{Z}_p^*.$$

The public key is the vector

$$(p, n; u_0, \ldots, u_n) \in \mathbf{Z}^2 \times (\mathbf{Z}_p^*)^{n+1}$$

and the secret key is s.

(ii) Encryption: Let m be a message and $\sum_{i=0}^{n} 2^i m_i$ its binary expansion. The encryption of the $n+1$ bit message m is $c = \prod_{i=0}^{n} u_i^{m_i} \bmod p$.

(iii) Decryption: To decrypt the ciphertext c, we compute

$$m = \sum_{i=0}^{n} \frac{2^i}{p_i - 1} \times \big(\gcd(p_i, c^s \bmod p) - 1 \big).$$

From the description of the NSK scheme, we see that the security is based on the DLP. It is sufficient to solve $u_i^x = p_i$ in $\mathbf{Z}_p^*$ for some i. The best algorithm for computing DLP in prime fields has subexponential bit complexity [1,10]. Thus, for large p (at least 2048 bits), the system cannot be attacked by using the state of the art algorithms for DLP.

We have also assumed that the prime number p belongs to the special class of safe primes to prevent attacks, such as Pollard rho [28], Pollard $p - 1$ algorithm [27], Pohlig–Hellman algorithm [26] or any similar procedure that exploits properties of $p - 1$.

4.2.   *The attack*

Since $c \equiv \sum_{i=0}^{n} 2^i m_i \bmod \Lambda$, we get

$$\prod_{i \in I_1} u_i^{m_i} = c \prod_{i \in I_2} u_i^{-m_i} \quad \text{(in } \mathbf{Z}_\Lambda\text{)}. \tag{9}$$

So, we can apply BA_MSPP_Λ with input $\mathcal{P} = (u_i)_i$ and c and for some bound b and Hamming weight of the message m, say H_m, i.e. the number of 1s in the binary message m. So, in this attack, we assume that we know the Hamming weight or an upper bound of it. To be more precise, this attack is feasible only for small or large Hamming weights. Our parallel version allows us to consider larger Hamming weights than in Ref. [4].

In the following algorithm, we call Algorithm 2, where we execute steps 4 and 5 in parallel (in function BA_MSPP_Λ).

Algorithm 3. Attack to NSK cryptosystem

INPUT: ∘ The cryptographic message c
∘ the Hamming weight H_m of the message m
∘ a bound $b \leq n/2$
∘ the public key $pk = (p, n; u_0, \ldots, u_n)$ of NSK cryptosystem

OUTPUT: the message m or Fail

1: $\mathcal{P} \leftarrow \{u_0, \ldots, u_n\}$
2: $\mathbb{S} \leftarrow \text{BA\_MSPP}_p(\mathcal{P}, c; b, H_m)$
3: if $\mathbb{S} \neq \emptyset$ construct m. Else return Fail

4.2.1. Reduction of the case of large Hamming weight messages

The case where we have large Hamming weight of a message can be reduced to the case where we have small Hamming weight. Indeed, if the message m has $H_m = n+1-\varepsilon$, where ε is a small positive integer, then again we can reduce the problem to one with small Hamming weight. Let $c = \text{Enc}(m)$ and $c' = c^{-1}u_n^2 \prod_{i=0}^{n-1} u_i$. We provide the following lemma.

Lemma 4. *The decryption of c' is $m' = 2^{n+1} + 2^n - m - 1$, where $H_{m'} = \varepsilon + 1$.*

Proof. See Ref. [4, Lemma 3.4]. □

So, we can consider $H_m < b$. Indeed, if $H_m \geq b$, we apply the previous lemma and we get $H_{m'} < b$. So, finding m' is equivalent finding m.

4.2.2. The case of knowing some bits of the message

If we know the position of some bits of the message m (for instance, by applying a fault attack to the system may leak some bits), then we can improve our attack. In this case, we choose I_1 and I_2 in Algorithm 3, from the set $\{0, 1, \ldots, n\} - K$, where the set K contains the positions of the known bits. Also, in line 10, when we reconstruct the message m (in case of a collision) we need to put the known bits to the right positions.

4.3. Experimental results for NSK cryptosystem

In our implementation,[3] we used C/C++ with GMP library [12] and for parallelization OpenMP [25]. We used 20 threads of an Intel(R) Xeon(R) CPU E5-2630 v4 @2.20 GHz in a Linux platform.

First, in the following Table 2, we present the improvement of the results provided in Ref. [4] by using the parallel version.

Besides, in the following Table 3 we extend the results of the previous table. In fact, Table 3 demonstrates that having a suitable

---

[3]The code can be found https://goo.gl/t9Fa68.

Table 2. We used $b = n/2, Q = 12$, where $n = 84, 130$ and 232, for $\text{len}(p) = 600, 1024$ and 2048 bits, respectively. For each column, we executed 10 times the attack of [4] and our parallel version (Algorithm 3), and we computed the average CPU time.

$\text{len}(p)$	600			1024			2048	
H_m	8	9	10	8	9	10	7	8
Attack [4]	42 s	300 s	7.5 m	8.1 m	1.1 h	1.96 h	1.17 h	2.15 h
Parallel attack	<1 s	6.54 s	15.56 s	11.44 s	70 s	284 s	70 s	11.46 m

Table 3. Extension of Table 2. For the case $H_m = 13$, $Q = 12$, and $\text{len}(p) = 1024$ we used $b = 60$ instead of $b = \frac{n}{2} = 65$. For the case $H_m = 11$ and $\text{len}(p) = 2048$ we used $b = n/2 - 23 = 93$. For all other cases we used $b = \frac{n}{2}$. The last row, Average Rounds, is the (average) round that eventually our algorithm terminates. Theoretically is approximately $\frac{1}{\mathbb{P}}$ (see Lemma 1).

$\text{len}(p)$	600			1024			2048	
n		84			130			232	
H_m	12	13	14	11	12	**13**	9	10	**11**
Parallel (time)	4 m	7 m	13 m	26 m	66 m	142 m	52 m	107 m	895 m
Mem. (GB)	1.43	4.93	7.5	14.28	21.67	67.41	26.9	40.61	127.53
Average rounds	4.2	2.4	2.6	8.2	2.6	5.6	7.4	2.4	10

number of threads and considering a suitable bound b, we get a practical attack for low Hamming weight messages.

In Figure 2, we represent some of our data graphically.

4.4. *Carmichael numbers*

Fermat proved that if p is a prime number, then p divides $a^p - a$ for every integer a. This is known as Fermat's Little Theorem. The question if the converse is true has negative answer. In fact, in 1910, Carmichael noted that 561 provides such a counterexample. A Carmichael number[4] is a positive composite integer n such that $a^{n-1} \equiv 1 \pmod{n}$ for every integer a, with $1 < a < n$ and $\gcd(a, n) = 1$. They are named after Robert Daniel Carmichael (1879–1967). Although the Carmichael numbers between 561 and

---

[4]See also http://oeis.org/A002997.

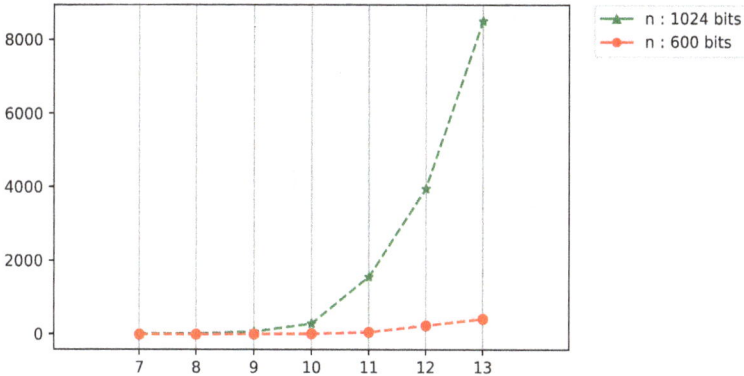

Figure 2.   The horizontal axis is the Hamming weight H_m and the vertical axis is the CPU time in seconds (for the parallel attack). Using **FindFit** of Mathematica [32], we computed the following approximation formulas that best fit to our data, $T_{600}(H_m) = 0.003 \cdot e^{0.71 H_m}$, $T_{1024}(H_m) = 0.029 \cdot e^{0.76 H_m}$ and $T_{2048}(H_m) = 0.74 \cdot e^{0.9 H_m}$ (seconds).

8911, i.e. the first seven, initially they were discovered by the Czech mathematician Šimerka in 1885 [31]. In 1910, Carmichael conjectured that there is an infinite number of Carmichael numbers. This conjecture was proved in 1994 by Alford, Granville, and Pomerance [2] although the problem of infinitely many Carmichael numbers with exactly $R \geq 3$ prime factors remains open until today. We have the following criterion.

Proposition 2 ([18]). *A positive integer n is Carmichael if and only if it is composite, square-free and for every prime p with $p|n$, we get $p - 1|n - 1$.*

For a simple elegant proof, see Ref. [6]. We define the following function:

$$\lambda(2^a) = \phi(2^a) \quad \text{if } a = 0, 1, 2$$

and

$$\lambda(2^a) = \frac{1}{2}\phi(2^a), \quad \text{if } a > 2.$$

If p_i is odd prime and a_i positive integer,

$$\lambda(p_i^{a_i}) = \phi(p_i^{a_i}).$$

If $n = p_1^{a_1} p_2^{a_2} \cdots p_k^{a_k}$, then

$$\lambda(n) = \mathrm{lcm}\left(\lambda(p_1^{a_1}), \lambda(p_2^{a_2}), \ldots, \lambda(p_k^{a_k})\right).$$

Korselt's criterion can be written as

Proposition 3 ([5]). *n is Carmichael if and only if it is composite and* $n \equiv 1 \pmod{\lambda(n)}$.

Using the previous proposition, we can prove that a Carmichael number is odd and it has at least three prime factors. Furthermore, we can calculate some Carmichael numbers (the first 16): 561, 1105, 1729, 2465, 2821, 6601, 8911, 10585, 15841, 29341, 41041, 46657, 52633, 62745, 63973 and 75361. In Ref. [3] they used an idea of Erdős [7] to find Carmichael numbers with many prime factors. In 1996, Loh and Niebuhr [19] provided a Carmichael with $1, 101, 518$ prime factors using Erdős heuristic algorithm (Algorithm 1). Also, an analysis and some refinements of Ref. [19] and an extension to other pseudoprimes were provided by Guillaume and Morain in 1996 in Ref. [11]. In the same chapter, the authors provided a Carmichael number having 5104 prime factors. In 2014 see Ref. [3, Table 1], the authors provided two large Carmichael numbers with many prime factors: the first one with $1, 021, 449, 117$ prime factors and $d = 25, 564, 327, 388$ decimal digits and the second with $10, 333, 229, 505$ prime factors, with $d = 29, 548, 676, 178$ decimal digits. Also, in 1975, Swift [30] generated all the Carmichael numbers below 10^9. In 1979, Yorinaga [34] provided a table for Carmichael numbers up to 10^{10} using a method of Chernick (see Ref. [11, Theorem 2.2]). In 1980, Pomerance, Selfridge and Wagstaff [29] generated Carmichael numbers up to $25 \cdot 10^{10}$. In 1988, Keller [17] calculated the Carmichael numbers up to 10^{13}. In 1990, Jaeschke [14] provided tables for Carmichael numbers up to 10^{12}. Pinch provided a table for all Carmichael numbers up to 10^{18} [21]. Also, the same author in 2006 [22] computed all Carmichael numbers up to 10^{20} and in 2007 [23] a table up to 10^{21}. Furthermore, he found 20138200 Carmichael numbers up to 10^{21} and all of them have at most 12 prime factors.

For an illustration of our algorithm we also generated some tables for Carmichael numbers having many prime factors.[5] For instance, we

---

[5]See https://github.com/drazioti/Carmichael.

produced Carmichael numbers up to 250 prime factors. Each instance was generated in some seconds. Also, Carmichael numbers with 11725 and 19589 prime factors were generated in some hours with our algorithm in a small home PC (I3/16 Gbyte) using a C++/OpenMP implementation. The following method is based on the Erdős idea [7]. It was used in Refs. [3,19] to produce Carmichael numbers having large number of prime factors.

Algorithm 4. Generation of Carmichael numbers

`INPUT:`

A positive integer r and a vector $\mathbf{H} = (h_1, h_2, \ldots, h_r) \in \mathbf{Z}^r$, with $h_1 \geq h_2 \geq \cdots \geq h_r \geq 1$. Also, we consider two positive integers ℓ and b which correspond to the local Hamming and the bound, respectively.
`OUTPUT:` A Carmichael number or Fail

1: $Q \leftarrow \{q_1, \ldots, q_r\}$ the r-first prime numbers
2: $\Lambda \leftarrow q_1^{h_1} \cdots q_r^{h_r}$
3: $\mathcal{P} \leftarrow \{p : p \text{ prime } p - 1 | \Lambda, \ p \nmid \Lambda\}$
4: $S \leftarrow \text{BA\_MSPP}_\Lambda(\mathcal{P}, 1; b, \ell)$
5: **If** $|S| \geq 2$ **return** $\prod_{p \in S} p$
6: $B \leftarrow \prod_{p \in \mathcal{P}} p$
7: $T \leftarrow \text{BA\_MSPP}_\Lambda(\mathcal{P}, B; b, \ell)$
8: **If** $|T| \geq 2$ **return** $B / \prod_{p \in T} p$
9: **else return** Fail

In line 8, we return the number $\prod_{p \in \mathcal{P} - T} p$.

Correctness: It is enough to prove that the numbers returned in steps 5 and 8 are Carmichael. Set $n = \prod_{p \in S} p$. We shall prove it for step 5. The set S contains all the primes of $\mathcal{P}$ such that their product is equivalent to 1 (mod Λ). Since $|S| \geq 2$, n is composite and also is square-free. Say a prime p is such that $p|n$. Since $n \equiv 1$ (mod Λ), i.e. $\Lambda|n - 1$, we get $p - 1|n - 1$. Indeed, this is immediate since $p - 1|\Lambda$. From Korselt's criterion, we get that n is Carmichael and similarly for step 8.

We have set $\ell = \text{local\_hamming}$. In case of success, the output of the algorithm is a Carmichael number with ℓ or $|\mathcal{P}| - \ell$ prime factors. In fact, if we want to calculate a Carmichael number with many prime factors, we can ignore lines 4 and 5 and consider a large set $\mathcal{P}$. An estimation for $|\mathcal{P}|$ was given in Ref. [19, formula 4]:

$$|\mathcal{P}| \approx g(\Lambda) \prod_{j=1}^{r} \left(h_j + \frac{q_j - 2}{q_j - 1} \right), \quad \text{where } g(\Lambda) = \frac{\Lambda}{\phi(\Lambda) \ln \sqrt{2\Lambda}}.$$

In lines 1 and 2, we initialize the algorithm. Since in practice r is not large enough, both these steps are very efficient.

In line 3, we calculate the set $\mathcal{P}$. One way to construct this set is the following. Say $d | \Lambda$. If $d + 1$ is prime with $d + 1 \notin Q$, then $d \in \mathcal{P}$. Finding the divisors of Λ having their prime divisors is a simple combinatorial problem. We can implement this without using much memory. Even better, we can use Ref. [3, Section 8], where they keep only the exponents of the divisors of Λ. Since the set $\mathcal{P}$ contains integers of the form

$$2^{a_1} 3^{a_2} \cdots p_r^{a_r} + 1,$$

with $0 \le a_i \le h_i$, instead of storing

$$2^{a_1} 3^{a_2} \cdots p_r^{a_r} + 1,$$

we can store $(a_1, \ldots, a_r)$. Overall, $8r|\mathcal{P}|$ bits or $r|\mathcal{P}|$ bytes. So, the set $\mathcal{P}$ is *nice*, since the set B in formula (4) needs $O(|\mathcal{P}| \log_2(|\mathcal{P}|))$ bits for storage.

In line 4 (and 7), we use Algorithm 2 with $b = $ bound and $\ell = $ local_hamming according to the user choice. We can apply BA_MSPP with the parameters Q and iter, BA_MSPP$_\Lambda(\mathcal{P}, c; b, \ell, Q, \text{iter})$. In Ref. [19], they picked T randomly from $\mathcal{P}$.

Remark 3. In Ref. [3], they used another algorithm inspired by the quantum algorithm of Kuperberg and they exploit the distribution of the primes in the set $\mathcal{P}$ (which is not uniform).

Remark 4. When $|\mathcal{P}|$ is large enough, then using $B = 1$ as target number, we can easily find a Carmichael number with small number of prime factors (by using small local Hamming weight). If we use $B > 1$ as in line 5, we get a Carmichael number with many prime factors. As we remarked earlier, the number of prime factors of the

Carmichael number is either ℓ or $|\mathcal{P}| - \ell$. One advantage of the algorithm is that we can search for Carmichael numbers near $|\mathcal{P}| - r$. This can be done by considering $\ell = $ local_hamming close to r. In this way, we quickly generated Carmichael numbers up to 250 prime factors in a small PC.

5. Conclusion

In this work, we considered a parallel algorithm to attack the modular version of product subset problem. This is an NP-complete problem which have many applications in computer science and mathematics. Here, we provide two applications: one in number theory and the other to cryptography.

First, we applied our algorithm (providing a C++ implementation) to the problem of searching Carmichael numbers. We managed to find one with 19589 factors in a small PC in 3 hours.

For the NSK cryptosystem, we updated and extended previous experimental cryptanalytic results provided in Ref. [4]. The new bounds for H_m concern messages having Hamming weight ≤ 11 or ≥ 223, for $n = 232$. This is proved by providing experiments. But, our attack is feasible for Hamming weight ≤ 15 or ≥ 219. The NSK cryptosystem system could resist to this attack if we consider Hamming weights in the real interval $[17, 217]$.

Acknowledgment

The authors are grateful to High Performance Computing Infrastructure and Resources (HPC) of the Aristotle University of Thessaloniki (AUTH, Greece) for providing access to their computing facilities and their technical support.

References

[1] L. Adleman (1979). Subexponential algorithm for the discrete logarithm problem with applications to cryptography, *SFCS '79 Proceedings of the 20th Annual Symposium on Foundations of Computer Science*, doi: 10.1109/SFCS.1979.2.

[2] W. R. Alford, A. Granville, and C. Pomerance (1994). There are infinitely many Carmichael numbers. *Ann. of Math. (2)*, **139**(3), 703–722.

[3] W. R. Alford, J. Grantham, S. Hayman, and A. Shallue (2014). Constructing Carmichael numbers through improved subset-product algorithm. *Math. Comp.*, **83**, 899–915.

[4] M. Anastasiadis, N. Chatzis, and K. A. Draziotis (2018). Birthday type attacks to the Naccache–Stern knapsack cryptosystem. *Inform. Proc. Lett.*, **138**, 39–43.

[5] R. Carmichael (1912). On composite numbers P which satisfy the Fermat congruence $a^{P-1} \equiv 1 \bmod P$. *Amer. Math. Monthly*, **19**(2), 22–27.

[6] J. Chernick (1939). On Fermat's simple theorem, *Bull. Amer. Math. Soc.*, **45**(4), 269–274.

[7] P. Erdős (1956). On pseudoprimes and Carmichael numbers. *Publ. Math. Debrecen*, **4**, 201–206.

[8] M. Fellows and N. Koblitz (1993). Fixed-parameter complexity and cryptography. *Applied Algebra, Algebraic Algorithms and Error-Correcting Codes* (Springer, Berlin, Heidelberg), pp. 121–131.

[9] M. R. Garey and D. S. Johnson (1979). *A Guide to the Theory of NP-Completeness* (W.H. Freeman, New York, NY).

[10] D. Gordon (1993). Discrete logarithms in $GF(p)$ using the number field sieve. *SIAM J. Discrete Math.*, **6**(1), 124–138.

[11] D. Guillaume and F. Morain (1996). Building pseudoprimes with a large number of prime factors. *Appl. Algebra Engrg. Comm. Comput.*, **7**(4), 263–277.

[12] T. Granlund and the GMP development team: GNU MP, The GNU Multiple Precision Arithmetic Library (2018).

[13] D. Harvey and J. V. D. Hoeven (2019). Integer multiplication in time $O(n \log n)$. https://hal.archives-ouvertes.fr/hal-02070778/document.

[14] G. Jaeschke (1990). Carmichael numbers to 10^{12}. *Math. Comput.*, **55**(191), 383–389.

[15] A. Joux (2009). *Algorithmic Cryptanalysis* (CRC Press).

[16] A. Karatsuba and Y. Ofman (1963). Multiplication of multidigit numbers on automata. *Sov. Phys. Dokl.* **7**, 595–596.

[17] W. Keller (1988). The Carmichael numbers to 10^{13}. *Abstracts of Papers Presented to the American Mathematical Society*, Vol. **9**, (American Mathematical Society) pp. 328–329, Abstract 88T-11-150.

[18] A. Korselt (1899). Probléme chinois', *Intermé. Math.*, **6**, 14–143.

[19] G. Loh and W. Niebuhr (1996). A new algorithm for constructing large Carmichael numbers. *Math. Comp.*, **65**(214), 823–836.

[20] G. Micheli, J. Rosenthal and R. Schnyder (1996). An information rate improvement for a polynomial variant of the Naccache–Stern knapsack cryptosystem. In *Physical and Data-link Security Techniques for Future Communication System*. Lecture Notes in Electrical Engineering, Vol. 358 (Springer, Cham), pp. 173–180.

[21] R. G. E. Pinch (1996). The Carmichael numbers up to 10^{18}. https://arxiv.org/pdf/math/0604376.pdf.

[22] R. G. E. Pinch (2006). The Carmichael numbers up to 10^{19}. ANTS, Berlin.

[23] R. G. E. Pinch (2007). The Carmichael numbers up to 10^{21}. *Conference on Algorithmic Number Theory Turku*. http://www.s369624816.websitehome.co.uk/rgep/p82.pdf.

[24] D. Naccache and J. Stern (1997). A new public key cryptosystem. *International Conference on the Theory and Applications of Cryptographic Techniques*. EUROCRYPT '97, Lecture Notes in Computer Science Vol. 1233 (Springer, Berlin, Heidelberg), pp. 27–36.

[25] OpenMP Architecture Review Board: Openmp application program interface version 3.0. http://www.openmp.org/mp-documents/spec30.pdf.

[26] S. Pohlig and M. Hellman (1978). An improved algorithm for computing logarithms over $GF(p)$ and its cryptographic significance. *IEEE Trans. Inform. Theory*, **24**, 106–110.

[27] J. M. Pollard (1974). Theorems of factorization and primality testing. *Proc. Cambridge Philos. Soc.*, **76**(3), 521–528.

[28] J. M. Pollard (1975). A Monte Carlo method for factorization. *Numer. Math.*, **15**(3), 331–334.

[29] C. Pomerance, J. L. Selfridge, and S. S. Wagstaff Jr. (1980), The pseudoprimes to $25 \cdot 10^{10}$, *Math. Comp.*, **35**(151), 1003–1026.

[30] J. D. Swift (1975). Review 13. *Math. Comp.*, **29**, 338–339.

[31] V. Šimerka (1885). Zbytky z arithmetické posloupnosti [On the remainders of an arithmetic progression]. *Cas. Pestovani Mat. Fys.*, **14**(5), 221–225.

[32] Wolfram Research Inc., (2018). Mathematica. https://www.wolfram.com/mathematica/.

[33] A. C. Yao (1978). New algorithms for bin packing. Report No. STAN-CS-78-662 Stanford University, Stanford, CA.

[34] M. Yorinaga (1979). Numerical computation of Carmichael numbers. II. *Math. J. Okayama Univ.*, **21**(2), 183–205.

Chapter 6

Cotangent Sums Related to Riemann Hypothesis

Mouloud Goubi

*Department of Mathematics, University Mouloud
Mammeri of Tizi-Ouzou 15000, Algeria
Algebra and Number Theory Laboratory, USTHB, Algiers, Algeria
Head of Laboratory of Pure and Applied Mathematics (LMPA),
Tizi-Ouzou*

mouloud.goubi@ummto.dz

Riemann zeta function plays an important role in number theory and can be used in the development of modern cryptography. The study of the Riemann hypothesis involves Vasyunin cotangent sums which are a generalization of the well-known Dedekind sums. The present survey chapter summarizes the various investigations on these sums and their law of reciprocity to apply the obtained results on Vasyunin formula linked to the Báez-Duarte criterion for the Riemann hypothesis.

1. Introduction

The study of Riemann zeta function has led to several approaches to attack the Riemann hypothesis. The most important is the Hilbertian approach initiated by Beurling–Nyman and developed in its modern form by Báez-Duarte and Balazard. The Vasyunin formula is an important tool to calculate the inner products of a certain family of vectors. In this formula, we see the Vasyunin cotangent sums which are a generalization of the Dedekind sums. In this survey chapter, we expose different investigations on Vasyunin cotangent sums and their reciprocity law in the spirit of improving the computation of the curious part of the Vasyunin formula.

2.   Cryptography

Modern cryptography [16] is focused on formal reasoning and analysis. We need formal definitions, precise assumptions and proofs of security. In public-key code, only the private key must be kept secret, and the public key may be shared. Modern cryptography needs knowledge in analytic and algebraic number theory. In each case, we give an example to better explain the strong link between cryptography and number theory. For public-key RSA, let ϕ be the Euler function and $M = pq$, where p, q are two random n-bit primes and we choose $a > 1$ such that $(a, \phi(M)) = 1$, thereafter we compute $\bar{a}$ the inverse of a modulo $\phi(M)$. The public key is (M, a) and the private key is $(M, \bar{a})$. Encryption and decryption for a message $m \in \{0, 1, 2, \ldots, M - 1\}$ are, respectively, $m^a \pmod{M}$ and $c^{\bar{a}} \pmod{N}$, with $c = m^a$. For verification, we have the identity $c^{\bar{a}} = (m^a)^{\bar{a}} = m$. The secret of this algorithm is the prime factorization of the modulus M. If the primes $p; q$ such that $M = pq$ are known, then the private key becomes public. Let $\mathbb{R}^n$ be the $\mathbb{R}$-vectorial space with the classical inner product. A lattice L is a subgroup of the form $L = \mathbb{Z}v_1 + \mathbb{Z}v_2 + \cdots + \mathbb{Z}v_m$ with $v_1, v_2, \ldots, v_m$ are linearly independent. Let the matrix $A = (\langle v_i, v_j \rangle)_{1 \leq i,j \leq m}$, then the volume of L is $\mathrm{vol}(L) = \sqrt{|\det A|}$. Lattice cryptography is based on the shortest vector problem: given a basis of a lattice L and find $v \in L$ such that $\|v\| = \min_{0 \neq x \in L} \|x\|$. Or we can fix $y \in \mathbb{R}^n$ and find the vector v such that $\|v - y\| = \min_{y \neq x \in L} \|x - y\|$. The Riemann zeta function $\zeta(s) = \sum_{n \geq 1} \frac{1}{n^s}$ is the first that we think of because according to the Euler formula, it is the product of numbers associated with the sequence of prime numbers:

$$\zeta(s) = \prod_{p \, \mathrm{prime}} \frac{1}{1 - p^{-s}}.$$

Riemann zeta function admits this natural generalization to Dirichlet series $L(s, \chi) = \sum_{n \geq 1} \frac{\chi(n)}{n^s}$ which are also written in the following way:

$$L(s, \chi) = \prod_{p \, \text{prime}} \frac{1}{1 - \chi(p)p^{-s}}.$$

3. Riemann Hypothesis and Vasyunin Cotangent Sums

The Riemann hypothesis predicts that all the zeros of zeta function have the real part equal to $\frac{1}{2}$. Let α be a positive real number and consider $e_\alpha = \{\frac{t}{\alpha}\}$, $t > 0$, where $\{t\} = t - \lfloor t \rfloor$ the fractional part function. These functions lie to the Hilbert space $\mathcal{H} = L^2\left(0, +\infty; t^{-2}dt\right)$. Nyman [23] showed that Riemann hypothesis is equivalent to the density on $\mathcal{H}$ of the subspace $\mathcal{B}$ generated by the functions $\alpha e_\alpha - e_1$, $\alpha \geq 1$. Báez-Duarte [4] proved that we can limit to α positive integers in this criterion. This returns back to show that χ the indicator function of the interval $[1, +\infty]$ is a limit on $\mathcal{H}$ of linear combinations of elements of $\mathcal{B}$ [4]. Let the distance

$$D(1, 2, \ldots, n) = \text{dist}_{\mathcal{H}}\left(\chi, \text{Vect}(e_1, e_2, \ldots)\right).$$

The Riemann hypothesis [3] is true if and only if $\lim_{n \to \infty} D(1, 2, \ldots, n) = 0$. Báez-Duarte *et al.* [5] conjectured that

$$d_n^2 \sim \frac{2 + \gamma - \log 4\pi + o(1)}{\log n}, \quad n \to +\infty, \tag{1}$$

and Balazard–de Roton [6] proved that

$$d_n^2 \geq \frac{2 + \gamma - \log 4\pi}{\log n}, \quad n \to +\infty. \tag{2}$$

It is well known that the conjecture (1) implies the Riemann hypothesis. A classical result of Hilbert geometry states that

$$D(e_1, e_2, \ldots, e_n)^2 = \frac{\text{Gram}(\chi, e_1, \ldots, e_n)}{\text{Gram}(e_1, \ldots, e_n)}, \tag{3}$$

where $\text{Gram}(u_1, \ldots, u_n) := \det\left(\langle u_j, u_k \rangle\right)_{1 \leq j, k \leq k}$ and $\langle u, v \rangle$ the inner product in $\mathcal{H}$ given by the expression

$$\langle u, v \rangle = \int_0^{+\infty} u(t)v(t)t^{-2}dt.$$

Two kinds of inner product appear in the formula (3); the first is

$$\langle \chi, e_p \rangle = \frac{\log p + 1 - \gamma}{p},$$

and the second, for $(p, q) = 1$, is given by the Vasyunin formula [25]:

$$\langle e_p, e_q \rangle = \frac{\log 2\pi - \gamma}{2} \left(\frac{1}{p} + \frac{1}{q} \right)$$

$$+ \frac{p - q}{2pq} \log \frac{q}{p} - \frac{\pi}{2pq} \left(V\left(\frac{q}{p}\right) + V\left(\frac{p}{q}\right) \right), \qquad (4)$$

which can be written under the form [11]

$$\frac{1}{2\pi\sqrt{pq}} \int_{-\infty}^{+\infty} \left| \zeta\left(\frac{1}{2} + it\right) \right| \left(\frac{q}{p}\right)^{it} \frac{dt}{\frac{1}{4} + t^2}$$

$$= \frac{\log 2\pi - \gamma}{2} \left(\frac{1}{p} + \frac{1}{q} \right)$$

$$+ \frac{p - q}{2pq} \log \frac{q}{p} - \frac{\pi}{2pq} \left(V\left(\frac{q}{p}\right) + V\left(\frac{p}{q}\right) \right).$$

$V\left(\frac{p}{q}\right)$ is the so-called Vasyunin cotangent sum [25] given by the relation

$$V\left(\frac{p}{q}\right) = \sum_{k=1}^{q-1} \left\{ \frac{kp}{q} \right\} \cot\left(\frac{\pi k}{q}\right).$$

It is connected to cotangent sum [18,21]

$$c_0\left(\frac{p}{q}\right) = \sum_{k=1}^{q-1} \left\{ \frac{k}{q} \right\} \cot\left(\frac{\pi kp}{q}\right)$$

by virtue of the relation

$$V\left(\frac{p}{q}\right) = -c_0\left(\frac{\bar{p}}{q}\right),$$

where $\bar{p}$ is the inverse of p modulo q. $c_0\left(\frac{p}{q}\right)$ appears on Estermann zeta function

$$E\left(s, \frac{p}{q}, \alpha\right) = \sum_{n \geq 1} \frac{\sigma_\alpha(n) \exp\left(2\pi \, in \, p/q\right)}{n^s}, \qquad (5)$$

at $s = 0$, more precisely, we have

$$E\left(0, \frac{p}{q}, 0\right) = \frac{1}{4} + \frac{i}{2}c_0\left(\frac{p}{q}\right). \tag{6}$$

Bettin–Conrey [12] proved the following reciprocity formula for the $c_0\left(q/p\right)$:

$$c_0\left(\frac{q}{p}\right) + \frac{p}{q}c_0\left(\frac{p}{q}\right) - \frac{1}{\pi q} = \frac{i}{2}\psi_0\left(\frac{q}{p}\right),$$

where

$$\psi_0\left(z\right) = -2\frac{\log 2\pi z - \gamma}{\pi i z} - \frac{2}{\pi}\int_{\left(\frac{1}{2}\right)}\frac{\zeta(s)\zeta(1-s)}{\sin \pi s}z^{-s}ds.$$

The similar reciprocity formula for Vasyunin cotangent sums [13] is

$$\frac{\bar{q}}{p}V\left(\frac{q}{p}\right) + V\left(\frac{\bar{p}}{\bar{q}}\right) + \frac{1}{\pi p} = -g\left(\frac{\bar{q}}{p}\right), \tag{7}$$

where $g\left(\frac{\bar{q}}{p}\right) = \frac{i\bar{q}}{2p}\psi_0\left(\frac{\bar{q}}{p}\right)$ is an analytic function in $\mathbb{C}\backslash\mathbb{R}_-$, which has the following asymptotic expansion of order $N \geq 2$ and $x \to 0$:

$$g\left(x\right) = -\frac{\log 2\pi x - \gamma}{\pi} + \frac{2}{\pi}\sum_{k=2}^{N}\frac{\zeta\left(k\right)B_k}{k}x^k + O\left(x^{N+1}\right).$$

For a good approach of the Báez-Duarte criterion, we need a formula for the symmetric sum $S(p,q) = V\left(\frac{q}{p}\right)+V\left(\frac{p}{q}\right)$. In Ref. [7], we showed that

$$S\left(p,q\right) = \frac{1}{\pi}\left(G\left(p,p\right) + G\left(q,q\right) - 2G\left(p,q\right) + (q-p)\log\frac{p}{q}\right), \tag{8}$$

where

$$G\left(p,q\right) = \sum_{k\geq 1}\frac{pq}{k\left(k+1\right)}\left\{\frac{k}{p}\right\}\left\{\frac{k}{q}\right\}. \tag{9}$$

A natural generalization of the Vasyunin cotangent sums [7] is defined by

$$V_a\left(\frac{p}{q}\right) = -q^a\sum_{k=1}^{q-1}\zeta\left(-a, \left\{\frac{kp}{q}\right\}\right)\cot\left(\frac{\pi k}{q}\right), \tag{10}$$

where $a \in \mathbb{Z}$ and $\zeta(s, x)$ is the well-known Hurwitz zeta function [2] given by the series

$$\zeta(s, x) = \sum_{k \geq 0} \frac{1}{(k+x)^s}, \quad \Re(s) > 1.$$

$V_a\left(\frac{p}{q}\right)$ is connected to the cotangent sum [11]

$$C_a\left(\frac{p}{q}\right) = q^a \sum_{k=1}^{q-1} \cot\left(\frac{\pi k p}{q}\right) \zeta\left(-a, \frac{k}{q}\right) \tag{11}$$

by virtue of the relation

$$V_a\left(\frac{p}{q}\right) = -C_a\left(\frac{\bar{p}}{q}\right).$$

The reciprocity law satisfied by $V_a\left(\frac{\bar{p}}{q}\right)$ is given by the following sample form [7]:

$$\left(\frac{q}{p}\right)^{1+a} V_a\left(\frac{\bar{q}}{p}\right) - V_a\left(\frac{\bar{p}}{q}\right) = \frac{q^a B_a}{\pi p}. \tag{12}$$

To have a good evaluation of the distance $D(1, 2, \ldots, n)$, we investigate the sums $V(1/q)$, $V(q/p)$ and the symmetric sum $S(p, q)$ to explain the relationship between $V(1/q)$ and $V(p/q)$. We start by studying the sums $V(1/q)$ and $V_a(p/q)$ to end with the symmetric sum $S(p, q)$ which makes it possible to improve the curious part of the formula of Vasyunin.

4. Cotangent Sum $V(1/q)$

Vasyunin [25] proved in a short way the following asymptotic formulation:

$$V\left(\frac{1}{q}\right) = -\frac{1}{\pi} q \log q + \frac{q}{\pi}(\log 2\pi - \gamma) + O(\log q). \tag{13}$$

Rassias [24] improved the error term in the Vasyunin asymptotic formula (13), namely:

$$V\left(\frac{1}{q}\right) = -\frac{1}{\pi} q \log q + \frac{q}{\pi}(\log 2\pi - \gamma) + O(1) \tag{14}$$

and showed that

$$V(1/q) = -\frac{1}{\pi}q\log q + \frac{q}{\pi}(\log 2\pi - \gamma) + \frac{1}{\pi}$$

$$- \sum_{l=1}^{n} E_l q^{-l} - R_n^{\star}(q) \tag{15}$$

for chosen sequences E_l and R_l and the integer n such that $q \geq 6\left(\lfloor\frac{n}{2}\rfloor + 1\right)$. In addition, he established the series expansion

$$c_0\left(\frac{1}{q}\right) = \frac{1}{\pi}\sum_{\substack{a>1 \\ q\nmid a}}\left[\frac{q}{a}\left(1 + 2\left\lfloor\frac{a}{q}\right\rfloor\right) - 2\right]. \tag{16}$$

We can also recall this integral representation [13]

$$V(1/q) = \frac{1}{\pi}\int_0^1 \frac{(q-2)t^q - qt^{q-1} + qt - q + 2}{(1-t)^2(1-t^q)}dt. \tag{17}$$

It is easily checked that

$$(q-2)t^q - qt^{q-1} + qt - q + 2 = (t-1)^3\sum_{r=1}^{q-1}(q-r-1)\,rt^{r-1}$$

and

$$V(1/q) = \frac{1}{\pi}\int_0^1 \frac{\sum_{r=1}^{q-1}(q-r-1)\,rt^{r-1}}{1+t+\cdots+t^{q-1}}dt, \tag{18}$$

which can become

$$V(1/q) = \frac{1}{\pi}p\log p - \frac{1}{\pi}\int_0^1 \frac{\sum_{r=1}^{q-1}(r+1)\,rt^{r-1}}{1+t+\cdots+t^{q-1}}dt. \tag{19}$$

To succeed in transforming the integral (17) into a series, one must recall briefly the notion of generating functions; tools we need to find the series expansions of $V\left(\frac{1}{q}\right)$ and give another proof [15] of the Dirichlet series found by Rassias [24] based on arithmetical tools without using the Fourier analysis. Formally, $f(t)$ is a generating function of a sequence a_n of numbers if we have $f(t) = \sum_{n\geq0}a_n t^n$.

The Cauchy product of the functions $f(t)$ and $g(t) = \sum_{n\geq 0} b_n t^n$ is defined by means of the relation

$$f(t)g(t) = \sum_{n\geq 0} \left(\sum_{k=0}^{n} a_k b_{n-k} \right) t^n. \tag{20}$$

In this order of ideas, let the sequence ℓ_n defined by the generating function

$$\frac{2 - q + qt - qt^{q-1} + (q-2)t^q}{(1-t)^2 (1-t^q)} = \sum_{n\geq 0} \ell_n t^n, \quad |t| < 1. \tag{21}$$

From the integral representation (17), the Dirichlet series of $V(1/q)$ is

$$\pi V (1/q) = \pi \sum_{n\geq 0} \frac{\ell_n}{n+1}. \tag{22}$$

Let the sequence $b_n(q)$ be defined by the generating function

$$\frac{1}{(1-t^q)(1-t)^2} = \sum_{n\geq 0} b_n(q)t^n.$$

We have $b_0(q) = 1$, $b_1(1) = 2$ and the recursive formulas [10]:

$$\begin{cases} b_n(q) - 2b_{n-1}(q) + b_{n-2}(q) = 0, & 2 \leq n \leq q-1, \; n = q+1, \\ b_q(q) - 2b_{q-1}(q) + b_{q-2}(q) = 1, \\ b_n(q) - 2b_{n-1}(q) + b_{n-2}(q) - b_{n-q}(q) \\ \quad +2b_{n-q-1}(q) - b_{n-q-2}(q) = 0, & n \geq q+1. \end{cases}$$

For $t \in \,]-1,1[$, the series expansion of rational function $\frac{2-q+qt-qt^{q-1}+(q-2)t^q}{(1-t)^2(1-t^q)}$ is

$$\sum_{n\geq 0} b_n(q) \left[(2-q)t^n + qt^{n+1} - qt^{n+q-1} + (q-2)t^{n+q} \right].$$

The integral of this series on the interval $]-1, 1[$ leads to the following series expansion of $V(1/q)$:

$$V(1/q) = \frac{1}{\pi}q(q-1)(q-2) \sum_{n\geq 0} \frac{b_n(q)}{(n+1)(n+2)(n+p)(n+p+1)}.$$

Since

$$\frac{1}{(1-t)^2} = \sum_{n\geq 0}(n+1)t^n \quad \text{and} \quad \frac{1}{1-t^q} = \sum_{n\geq 0}t^{qt},$$

and by the Cauchy product of generating functions, we can easily show that [10]

$$b_n(q) = \left(n+1-\frac{q}{2}\left\lfloor\frac{n}{q}\right\rfloor\right)\left(\left\lfloor\frac{n}{q}\right\rfloor+1\right).$$

Usually, we write $n = iq + r$ and the quantity $b_n(q)$ becomes

$$b_{iq+r}(q) = \left(n+1-\frac{q}{2}i\right)(i+1).$$

In addition, if we consider the sequence c_n defined by $c_0 = 1 - q$, $c_1 = q$, $c_{q-1} = -q$, $c_q = q - 2$. According to the sequence c_k and $b_k(q)$, a new expression of ℓ_n is given by the relation

$$\ell_n = \sum_{k=0}^{n} c_k b_{n-k}(q),$$

which once n exceeds q can become as follows:

$$\ell_n = (2-q)b_n(q) + qb_{n-1}(q) - qb_{n-q+1}(q) + (q-2)b_{n-q}(q).$$

This formula remains insufficient, in what follows, we propose the simple calculation to find the explicit formula of ℓ_n. We have

$$(1-t^q)\left(\sum_{n\geq 0}\ell_n t^n\right) = \left(\sum_{n\geq 0}c_n t^n\right)\left(\sum_{n\geq 0}(n+1)t^n\right),$$

and for $n \leq q - 1$,

$$\ell_n = \sum_{k=0}^{n}(n-k+1)c_k,$$

but for $n \geq q$, we deduce that

$$\ell_n - \ell_{n-q} = \sum_{k=0}^{n}(n-k+1)c_k,$$

which means that

$$\ell_{n+q} - \ell_n = \sum_{k=0}^{n+q}(n + q - k + 1)c_k.$$

But it is easily checked that $\sum_{k=0}^{n+q}(n + q - k + 1)c_k = 0$, thereafter $\ell_{n+q} = \ell_n$, which means that ℓ_n is periodic of period q. So, it suffices to know the values of ℓ_n over the interval $[0, q-1]$ to deduce the others. In this case, we have

$$\ell_r = \begin{cases} 2(r+1) - q, & \text{if } r \leq q - 2, \\ 0, & \text{if } r = q - 1. \end{cases} \tag{23}$$

Let r be the remainder obtained when n is divided into q, then $0 \leq r \leq q - 1$ and $\ell_n = \ell_r$. Thereafter,

$$\pi V(1/q) = \sum_{k \geq 0} \sum_{r=1}^{q-1} \frac{2r - q}{kq + r}. \tag{24}$$

Since we have $r = n - q\lfloor \frac{n}{q} \rfloor$ and r is different from q in the last sum, we deduce that

$$\pi V(1/q) = \sum_{\substack{n \geq 1 \\ q \nmid n}} \frac{2\left(n - q\left\lfloor \frac{n}{q} \right\rfloor\right) - q}{n}.$$

4.1. Connection to digamma function and Bernoulli polynomials

Let B_1 be the first reduced Bernoulli polynomial

$$B_1(x) = \begin{cases} 0 & \text{if } x \in \mathbb{Z}, \\ \{x\} - \dfrac{1}{2} & \text{otherwise}, \end{cases} \tag{25}$$

and ψ the digamma function given by

$$\psi(z) = -\gamma - \frac{1}{z} + \sum_{k \geq 1}\left(\frac{1}{k} - \frac{1}{k + z}\right).$$

The function ψ [13] satisfies the identity

$$\psi\left(z_1\right) - \psi\left(z_2\right) = \int_0^{+\infty} \frac{e^{-z_2 t} - e^{-z_1 t}}{1 - e^{-t}} dt$$

which implies that

$$\psi\left(z_1\right) - \psi\left(z_2\right) = \int_0^1 \frac{t^{z_2-1} - t^{z_1-1}}{1 - t} dt$$

and

$$\psi\left(z_1\right) = -\gamma + \int_0^1 \frac{1 - t^{z_1-1}}{1 - t} dt. \tag{26}$$

Furthermore, $\psi\left(z_1\right)$ is negative in the interval $[-\infty, 1]$. If $z_1 = n$ is a positive integer,

$$\psi\left(n + 1\right) = -\gamma + H_n,$$

where

$$H_n = \sum_{j=1}^n \frac{1}{j}$$

is the harmonic number of order n, which can be written in the following form:

$$H_n = \int_0^1 \frac{1 - t^n}{1 - t} dt.$$

Finally, the reflection formula (see Ref. [1, Section 6.3.7]) which relates ψ to cotangent function is given by

$$\psi\left(1 - z_1\right) - \psi\left(z_1\right) = \pi \cot\left(\pi z_1\right).$$

Báez-Duarte *et al.* [2] showed that

$$V\left(\frac{p}{q}\right) = -\frac{2}{\pi} \sum_{r=1}^q B_1\left(\frac{rp}{q}\right) \psi\left(\frac{r}{q}\right). \tag{27}$$

In what follows, we revisit the proof of identity (27) for $p = 1$. First, we have [8]

$$\frac{\log q}{p} = \sum_{k \geq 1} \frac{1}{k(k+1)} \left\{ \frac{k}{q} \right\}, \tag{28}$$

this result is obtained in this following way:

$$\sum_{k \geq 1} \frac{1}{k(k+1)} \left\{ \frac{k}{q} \right\} = \frac{1}{q} \sum_{i \geq 0} \sum_{r=1}^{q-1} \frac{1}{(iq+r)(iq+r+1)}$$

and

$$\frac{1}{(iq+r)(iq+r+1)} = \frac{1}{iq+r} - \frac{1}{iq+r+1} = \int_0^1 \left(t^{iq+r-1} - t^{iq+r} \right) dt.$$

Then

$$\sum_{k \geq 1} \frac{1}{k(k+1)} \left\{ \frac{k}{q} \right\} = \frac{1}{q} \int_0^1 \frac{\sum_{r=1}^{q-1} r t^{r-1}}{1 + t + \cdots + t^{q-1}}.$$

In addition, we have the following identity [15]:

$$\psi \left(\frac{r+1}{q} \right) - \psi \left(\frac{r}{q} \right) = q \int_0^1 \frac{t^{r-1}}{1 + t + \cdots + t^{p-1}} dt.$$

As it can also follow from the identity (26), consequently [13]

$$\log p = \frac{1}{p} \sum_{r=1}^{p-1} r \left(\psi \left(\frac{r+1}{p} \right) - \psi \left(\frac{r}{p} \right) \right)$$

and

$$\sum_{r=1}^{p} \psi \left(\frac{r}{p} \right) = -\gamma p - p \log p.$$

Combining these identities with the integral representation (18) to get

$$V \left(\frac{1}{p} \right) = \frac{1}{\pi} \log p + \gamma \frac{p-1}{\pi p} - \frac{1}{\pi p} \sum_{r=1}^{p-1} (2r - p - 1) \psi \left(\frac{r}{p} \right),$$

then

$$V\left(\frac{1}{p}\right) = -\frac{1}{\pi p}\sum_{r=1}^{p-1}(2r-p)\,\psi\left(\frac{r}{p}\right).$$

But

$$2r-p = 2p\left(\frac{r}{p}-\frac{1}{2}\right) = 2pB_1\left(\frac{r}{p}\right).$$

Thus, we have

$$V\left(\frac{1}{p}\right) = -\frac{2}{\pi}\sum_{r=1}^{p}B_1\left(\frac{r}{p}\right)\psi\left(\frac{r}{p}\right).$$

5. Cotangent Sum $V\left(\frac{p}{q}\right)$ and Generalization

Many results on the sum $V\left(\frac{1}{q}\right)$ are obtained, the idea that comes at first glance is to write $V\left(\frac{p}{q}\right)$ as a function of $V\left(\frac{1}{q}\right)$ to take advantage of these results. Let the sum

$$Q\left(\frac{p}{q}\right) = \sum_{m=1}^{q-1}\cot\left(\frac{\pi mp}{q}\right)\left\lfloor\frac{mp}{q}\right\rfloor. \tag{29}$$

The link [19] between $V(p/q)$ and $V(1/q)$ is given by the relation

$$c_0\left(\frac{p}{q}\right) = \frac{1}{p}c_0\left(\frac{1}{q}\right) - \frac{1}{p}Q\left(\frac{p}{q}\right)$$

and $V\left(\frac{p}{q}\right)$ is connected to $V\left(\frac{1}{q}\right)$ by the relation

$$V\left(\frac{p}{q}\right) = \frac{1}{\bar{p}}V\left(\frac{1}{q}\right) + \frac{1}{\bar{p}}Q\left(\frac{\bar{p}}{q}\right), \tag{30}$$

which still remains a relation between p and its inverse $\bar{p}$ modulo q. We consider the sum

$$R\left(\frac{p}{q}\right) = \sum_{m=1}^{q-1}\left\lfloor\frac{mp}{q}\right\rfloor\psi\left(\frac{m}{q}\right).$$

Then a relation between and p and $\bar{p}$ which involves $Q\left(\frac{p}{q}\right)$ and $R\left(\frac{p}{q}\right)$ is developed in the following theorem.

M. Goubi

Theorem 1. *We have*

$$(1 - p\bar{p}) V\left(\frac{1}{q}\right) = \frac{1}{\pi}\left(q\log q + \gamma q - \gamma\right)(\bar{p} - 1)\,p$$

$$+ \frac{2p}{\pi} R\left(\frac{\bar{p}}{q}\right) - Q\left(\frac{p}{q}\right).$$

Proof. Using the identity (27), we write

$$V\left(\frac{p}{q}\right) = pV\left(\frac{1}{q}\right) + \frac{2}{\pi}\sum_{m=1}^{q-1}\left(\left\lfloor\frac{mp}{q}\right\rfloor - \frac{p-1}{2}\right)\psi\left(\frac{m}{q}\right)$$

and

$$V\left(\frac{p}{q}\right) = pV\left(\frac{1}{q}\right) + \frac{2}{\pi}\sum_{m=1}^{q-1}\left\lfloor\frac{mp}{q}\right\rfloor\psi\left(\frac{m}{q}\right) - \frac{p-1}{\pi}\sum_{m=1}^{q-1}\psi\left(\frac{m}{q}\right).$$

Then,

$$V\left(\frac{p}{q}\right) = pV\left(\frac{1}{q}\right) + \frac{2}{\pi}\sum_{m=1}^{q-1}\left\lfloor\frac{mp}{q}\right\rfloor\psi\left(\frac{m}{q}\right)$$

$$- \frac{p-1}{\pi}\left(\gamma - \gamma q - q\log q\right).$$

Combining this result with the identity (30), we obtain

$$\frac{1}{\bar{p}}Q\left(\frac{\bar{p}}{q}\right) - \left(p - \frac{1}{\bar{p}}\right)V\left(\frac{1}{q}\right) + \frac{p-1}{\pi}\left(\gamma - \gamma q - q\log q\right)$$

$$= \frac{2}{\pi}\sum_{m=1}^{q-1}\left\lfloor\frac{mp}{q}\right\rfloor\psi\left(\frac{m}{q}\right),$$

and the desired result holds. $\qquad\square$

5.1. *Supervision of cotangent sum c_0 and associated moments*

Recently, Maier–Rassias [20] studied the maximum of $\left|c_0\left(\frac{p}{q}\right)\right|$ for the value of p in short interval $]A_0 q, (A_0 + \Delta)\,b[$ and proved for $0 < A_0 <$

$1, 0 < C < \frac{1}{2}, \Delta = q^{-C}$ and $0 < D < \frac{1}{2} - C$ that

$$\max_{A_0 b < r < (A_0 + \Delta)b} \left| c_0 \left(\frac{r}{b} \right) \right| \le \frac{D}{\pi} b \log b. \tag{31}$$

But in the interval $[1, q-1]$, we have the following inequality.

Theorem 2. (see [14])

$$\max_{1 \le p \le q-1} \left| c_0 \left(\frac{p}{q} \right) \right| \le \frac{1}{\pi} \left(\frac{\gamma}{\log q} - \frac{\gamma}{q \log q} + 1 \right) (q-2) \log q. \tag{32}$$

Proof. For $t \ne 0$, we have

$$\overline{B}_1(t) = \{t\} - \frac{1}{2},$$

and B_1 satisfies for $1 \le p \le q-1$ and $q \ge 2$ the double inequality

$$\frac{2-q}{2q} \le \overline{B}_1 \left(\frac{mp}{q} \right) \le \frac{q-2}{2q}.$$

This result is due to the inequality $\frac{1}{q} \le \{ \frac{mp}{q} \} \le \frac{q-1}{q}$. Since $\psi \left(\frac{m}{q} \right)$ is a negative number for $1 \le m \le q-1$, according to formula (27) of cotangent sum $V \left(\frac{p}{q} \right)$, we conclude that

$$-\frac{2-q}{\pi q} \sum_{m=1}^{q-1} \psi \left(\frac{m}{q} \right) \le V \left(\frac{p}{q} \right) \le -\frac{q-2}{\pi q} \sum_{m=1}^{q-1} \psi \left(\frac{m}{q} \right),$$

but we have

$$\sum_{m=1}^{q-1} \psi \left(\frac{m}{q} \right) = \gamma - \gamma q - q \log q.$$

Then,

$$\left| V \left(\frac{p}{q} \right) \right| \le \frac{1}{\pi} \left(1 - \frac{2}{q} \right) \left(\frac{\gamma}{\log q} - \frac{\gamma}{q \log q} + 1 \right) q \log q.$$

The last inequality is independent of p, then we can replace p by $\overline{p}$ and get the desired result (32), Theorem 2. This result is valid for every value of q not forcedly large. As example for $q = 2$, we deduce the well-known result $c_0 \left(\frac{1}{2} \right) = 0$. $\qquad \square$

The function $h(t) = \frac{1}{\pi}\left(1 - \frac{2}{t}\right)\left(\frac{\gamma}{\log t} - \frac{\gamma}{t\log t} + 1\right)$ is well defined in the interval $[2, +\infty[$ and satisfies the inequality $0 \le h(t) \le 1$. An improvement of result (31) is illustrated in the following corollary.

Corollary 1. *For $q \ge 3$ and every $1 \le p \le q - 1$, we have*

$$\left| c_0\left(\frac{p}{q}\right) \right| \le \frac{D'}{\pi} q \log q, \tag{33}$$

where $0 < D' < 1$ and exactly

$$D' = \frac{q-2}{q}\left(\frac{\gamma}{\log q} - \frac{\gamma}{q\log q} + 1\right).$$

Let the arithmetical function φ_q defined by

$$\varphi_q(m) = \sum_{(p,q)=1} \left\{\frac{mp}{q}\right\},$$

φ_q is periodic with period q and vanish in the set $q\mathbb{N}$. If q is a prime number, φ_q coincide with Euler function in $\mathbb{N}\backslash q\mathbb{N}$ and we have $\varphi_q(m) = \frac{1}{2}\varphi(q)$. According to the expression of φ_q, the following theorem holds true.

Theorem 3. *The partial sum of c_0 satisfies the following identity:*

$$\sum_{(p,q)=1} c_0\left(\frac{p}{q}\right) = \frac{2}{\pi}\sum_{m=1}^{q-1} \varphi_q(m)\psi\left(\frac{m}{q}\right)$$

$$+ \frac{1}{\pi}\varphi(q)\left(q\log q + \gamma q - \gamma\right). \tag{34}$$

If q is a prime number, then

$$\sum_{p=1}^{q-1} c_0\left(\frac{p}{q}\right) = 0. \tag{35}$$

Proof. From the relation between c_0 and V, we can deduce that

$$\sum_{(p,q)=1} c_0\left(\frac{p}{q}\right) = -\sum_{(p,q)=1} V\left(\frac{\overline{p}}{q}\right) = -\sum_{(p,q)=1} V\left(\frac{p}{q}\right)$$

and

$$\sum_{(p,q)=1} c_0\left(\frac{p}{q}\right) = \frac{2}{\pi} \sum_{m=1}^{q-1} \left(\sum_{(p,q)=1} \overline{B}_1\left(\frac{mp}{q}\right)\right) \psi\left(\frac{m}{q}\right).$$

But we know that

$$\sum_{(p,q)=1} \overline{B}_1\left(\frac{mp}{q}\right) = \varphi_q(m) - \frac{1}{2}\varphi(q).$$

Furthermore,

$$\sum_{(p,q)=1} c_0\left(\frac{p}{q}\right) = \frac{2}{\pi} \sum_{m=1}^{q-1} \varphi_q(m)\,\psi\left(\frac{m}{q}\right) - \frac{1}{\pi}\varphi(q) \sum_{m=1}^{q-1} \psi\left(\frac{m}{q}\right),$$

and the result (34), Theorem 3, follows. If q is prime, then $(m,q)=1$ for $1 \le m \le q-1$ and

$$\varphi_q(m) = \sum_{(p,q)=1} \left\{\frac{pm}{q}\right\} = \sum_{r=1}^{q-1} \frac{p}{q},$$

but it is well known that $\sum_{p=1}^{q-1} r = \frac{q(q-1)}{2}$, then $\varphi_q(m) = \frac{q-1}{2}$. Since the Euler function φ satisfies the identity $\varphi(q) = q-1$, the result (35) is deduced. $\qquad\square$

The restricted moment studied by Maier–Rassias satisfies the following universal inequality independent of parameter p.

Proposition 1.

$$\sum_{\substack{(p,q)=1 \\ A_0 q < r < A_1 q}} c_0\left(\frac{p}{q}\right)^{2k} \le \frac{A_1 - A_0}{\pi^{2k}} q^{2k+1} (\log q)^{2k}. \tag{36}$$

Proof. Using identity (33) Corollary 1, we obtain

$$c_0^{2k}\left(\frac{p}{q}\right) = \left|c_0\left(\frac{p}{q}\right)\right|^{2k} \le \frac{D'^{2k}}{\pi^{2k}}q^{2k}(\log q)^{2k}.$$

Then

$$\sum_{\substack{(p,q)=1 \\ A_0 q < r < A_1 b}} c_0\left(\frac{p}{q}\right)^{2k} \le \frac{(A_1 - A_0)\,q}{\pi^{2k}}q^{2k}(\log q)^{2k},$$

and the identity (36) follows. $\square$

5.2. *Continued fraction expansion of Vasyunin cotangent sums*

For real number x, let

$$x = a_0 + \cfrac{1}{a_1 + \cfrac{1}{a_2 + \cfrac{1}{a_3 + \cdots}}} = [a_0, a_1, a_3, \ldots]$$

be its continued fraction expansion with partial quotients $a_0 \in \mathbb{Z}$, $a_k \in \mathbb{N}\backslash\{0\}$ for $k \ge 1$. The reader is referred to Ref. [22, Section 7], for the elementary properties of continued fractions which will be used here. The coefficients a_i can be calculated by the algorithm:

(1) $x = \lfloor x \rfloor + \{x\}$, $a_0 = \lfloor x \rfloor$, $\xi_0 = \{x\}$, if $\xi_0 = 0$, then x is represented by $x = [a_0]$.

(2) If $\xi_0 \ne 0$, then $\lfloor \frac{1}{\xi_0} \rfloor > 1$, $r_1 = \frac{1}{\xi_0}$, we obtain $x = [a_0, r_1]$, $a_1 = \lfloor r_1 \rfloor$ and $\xi_1 = r_1 - a_1$; if $\xi_1 = 0$, then $x = a_0 + \frac{1}{r_1} = a_0 + \frac{1}{a_1} = [a_0, a_1]$.

(3) Otherwise, we take $r_2 = \frac{1}{\xi_1}$ and iterate the process.

The sequence $a_0, a_1, a_2, \ldots$ is finite if and only if x is a rational number. In the rational case, this algorithm is the Euclidean algorithm. Let us express a/b as a continued fraction of the form $a/b = [a_0, a_1, \ldots, a_n]$ and $a_{n+1} = 0$. With the Euclidean algorithm, we determine $a_0, a_1, \ldots, a_n$ by the recurrence

$$p_{k-1} = a_{k-1}p_k + p_{k+1}, \quad p_0 = a, p_1 = b, p_{n+1} = 0. \qquad (37)$$

Then $p_n = \gcd(p, q) = 1$. Let s_k, t_k be two sequences of integers

$$s_k = \begin{cases} 0 & \text{if } k = -2, \\ 1 & \text{if } k = -1, \\ a_k s_{k-1} + s_{k-2} & \text{if } k \geq 0, \end{cases}$$

$$t_k = \begin{cases} 1 & \text{if } k = -2, \\ 0 & \text{if } k = -1, \\ a_k t_{k-1} + t_{k-2} & \text{if } k \geq 0. \end{cases}$$

The sequences p_k, s_k and t_k satisfy the following properties [13]:

$$[a_0, a_1, \ldots, a_k] = \frac{s_k}{t_k}, \quad k \geq 0, \qquad (38)$$

$$[a_k, a_{k-1}, \ldots, a_1] = \frac{t_k}{t_{k-1}}, \quad k \geq 1, \qquad (39)$$

$$[0, a_{k-1}, \ldots, a_n] = \frac{p_k}{p_{k-1}}, \quad k \geq 0, \qquad (40)$$

$$[a_k, a_{k-1}, \ldots, a_0] = \frac{s_k}{s_{k-1}}, \quad k \geq 0, \qquad (41)$$

$$t_k s_{k-1} - t_{k-1} s_k = (-1)^k \qquad (42)$$

and

$$p_k = (-1)^k (a t_{k-2} - b s_{k-2}). \qquad (43)$$

Consequently, let $\frac{a}{b} = [a_0, a_1, \ldots, a_n]$, then we have

$$p_k = b \prod_{j=2}^{k} [0, a_{j-1}, \ldots, a_n], \quad k \geq 2 \qquad (44)$$

and

$$b = p_k t_{k-1} + p_{k+1} t_{k-2}, \quad k \geq 1. \qquad (45)$$

Following these properties, we can have

$$\sum_{k=1}^{n-1} \frac{(-1)^k}{p_k p_{k+1}} = \frac{1}{b^2 [a_0, a_1, \ldots, a_{n-2}] - ab}. \qquad (46)$$

The proof consists to write:

$$\sum_{k=1}^{n-1} \frac{(-1)^k}{p_k p_{k+1}} = \frac{1}{b} \sum_{k=1}^{n-1} \left(\frac{(-1)^k t_{k-1}}{p_{k+1}} + \frac{(-1)^k t_{k-2}}{p_k} \right)$$

$$= \frac{1}{b} \sum_{k=1}^{n-1} \frac{(-1)^k t_{k-1}}{p_{k+1}} + \frac{1}{b} \sum_{k=1}^{n-1} \frac{(-1)^k t_{k-2}}{p_k}$$

$$= \frac{1}{b} \sum_{k=2}^{n} \frac{(-1)^{k-1} t_{k-2}}{p_k} + \frac{1}{b} \sum_{k=1}^{n-1} \frac{(-1)^k t_{k-2}}{p_k}$$

$$= \frac{(-1)^{n-1} t_{n-2}}{p_n\, b} - \frac{t_{-1}}{b\, p_1}$$

$$= \frac{(-1)^{n-1} t_{n-2}}{p_n\, b}.$$

From (43), we deduce that

$$p_n = (-1)^n \left(a t_{n-2} - b s_{n-2} \right),$$

and then,

$$\frac{(-1)^{n-1} p_n}{t_{n-2}} = \left(b \frac{s_{n-2}}{t_{n-2}} - a \right). \tag{47}$$

The relation (38) implies that

$$\frac{s_{n-2}}{t_{n-2}} = [a_0, a_1, \ldots, a_{n-2}]. \tag{48}$$

From the relations (47) and (48), we have

$$\frac{(-1)^{n-1} t_{n-2}}{p_n b} = \frac{1}{b^2 [a_0, a_1, \ldots, a_{n-2}] - ab}.$$

Let $\bar{p}/q = [a_0, a_1, \ldots, a_n]$ be a finite continued fraction of $\bar{p}/q$ and the finite sequence $(p_k)_{0 \le k \le n+1}$ defined by $p_{k-1} = a_{k-1} p_k + p_{k+1}$ with $p_0 = \bar{p}$, $p_1 = q$ and $p_{n+1} = 0$. Then we have the following theorem.

Theorem 4.

$$V\left(\frac{p}{q}\right) = \frac{1}{\pi\left(q[a_0, a_1, \ldots, a_{n-2}] - \bar{p}\right)}$$

$$- \sum_{k=2}^{n} (-1)^k g\left([0, a_{k-1}, \ldots, a_n]\right) \prod_{j=2}^{k} [a_{j-1}, \ldots, a_n].$$

$$(49)$$

Proof. The proof is detailed in Ref. [13]. □

5.3. *Series expansion of Vasyunin cotangent sums*

According to reciprocity (7) and the expression of g, we write the following asymptotic development [13] of $V\left(\frac{\bar{a}}{qa+r}\right)$.

Theorem 5. *For $(a, q) = 1$ and $r < a$, we have*

$$V\left(\frac{\bar{a}}{qa+r}\right) = -\left(q + \frac{r}{a}\right) V\left(\bar{r}, a\right) - \frac{1}{\pi a}$$

$$+ \frac{1}{\pi}\left(q + \frac{r}{a}\right)\left(\log \frac{2\pi a}{qa+r} - \gamma\right)$$

$$+ \frac{1}{2} \sum_{k=1}^{\lfloor \frac{N}{2} \rfloor} (-1)^k \frac{4^k \pi^{2k-1} B_{2k}^2}{k\,(2k)!}\left(\frac{a}{qa+r}\right)^{2k-1}$$

$$+ O\left(\frac{1}{q^N}\right).$$

$$(50)$$

Proof. From reciprocity law (7), we have

$$\frac{a}{qa+r} V\left(\frac{\bar{a}}{qa+r}\right) + V\left(\frac{\bar{r}}{a}\right) = -\frac{1}{\pi\,(qa+r)} - g\left(\frac{a}{qa+r}\right),$$

and then, we obtain

$$V\left(\frac{\bar{a}}{qa+r}\right) = -\frac{qa+r}{a} V\left(\bar{r}, a\right) - \frac{1}{\pi a} - \frac{qa+r}{a} g\left(\frac{a}{qa+r}\right).$$

Moreover, for large q, we have

$$g\left(\frac{a}{qa+r}\right) = -\frac{1}{\pi}\left(\log\frac{2\pi a}{qa+r}-\gamma\right) - \frac{1}{2}\sum_{k=1}^{\lfloor\frac{N}{2}\rfloor}(-1)^k$$

$$\times\frac{4^k\pi^{2k-1}B_{2k}^2}{k\,(2k)!}\left(\frac{a}{qa+r}\right)^k + O\left(\frac{1}{q^{N+1}}\right).$$

$\square$

Taking $r = a - 1$, one remarks that for $a \geq 1$, $(q+1)a \equiv 1$ mod $(qa+a-1)$ and $(a-1)^2 \equiv 1\,(a)$. Then $\bar{a} = p+1$ and $\bar{r} = r = a-1$. Since $V(a-1,a) = V(1,a)$ and thanks to the relation (50), we obtain the relation

$$V(q+1, aq+a-1)$$

$$= -\left(q+1-\frac{1}{a}\right)V(1,a) - \frac{1}{\pi a} + \frac{1}{\pi}\left(q+1-\frac{1}{a}\right)$$

$$\times\left(\log\frac{2\pi a}{(q+1)a-1}-\gamma\right) + \frac{1}{2}\sum_{k=1}^{\lfloor\frac{N}{2}\rfloor}(-1)^k\frac{4^k\pi^{2k-1}B_{2k}^2}{k\,(2k)!}$$

$$\times\left(\frac{a}{(q+1)a-1}\right)^{2k-1} + O\left(\frac{1}{q^N}\right). \tag{51}$$

Finally from relation (51), we can get the asymptotic formula of $V(\frac{1}{q})$.

$$V\left(\frac{1}{q}\right) = \frac{1}{\pi}\left(\log\frac{2\pi}{q}-\gamma\right)q - \frac{1}{\pi} - \frac{\pi}{36q}$$

$$+ \frac{1}{2}\sum_{k=2}^{\lfloor\frac{N}{2}\rfloor}(-1)^k\frac{4^k\pi^{2k-1}B_{2k}^2}{k\,(2k)!}\left(\frac{1}{q}\right)^{2k-1} + O\left(\frac{1}{q^N}\right). \tag{52}$$

The relation (52) improved the asymptotic formulae (14) and (15) proved by Maier–Rassias in Ref. [18, Theorem 1.7]. In what follows, we give a few examples. For large q, we have

$$V(q+1, 2p+1) = -\frac{1}{2\pi} + \frac{1}{\pi}\left(q + \frac{1}{2}\right)\left(\log\frac{4\pi}{2p+1} - \gamma\right)$$

$$+ \frac{1}{2}\sum_{k=1}^{\lfloor\frac{N}{2}\rfloor}(-1)^k\frac{4^k\pi^{2k-1}B_{2k}^2}{k(2k)!}\left(\frac{2}{2q+1}\right)^{2k-1}$$

$$+ O\left(\frac{1}{q^N}\right),$$

$$V(2q+1, 3q+1) = -\frac{q+\frac{1}{3}}{3\sqrt{3}} - \frac{1}{3\pi} + \frac{1}{\pi}\left(q + \frac{1}{3}\right)\left(\log\frac{3\pi}{2} - \gamma\right)$$

$$+ \frac{1}{2}\sum_{k=1}^{\lfloor\frac{N}{2}\rfloor}(-1)^k\frac{4^k\pi^{2k-1}B_{2k}^2}{k(2k)!}\left(\frac{3}{3q+1}\right)^{2k-1}$$

$$+ O\left(\frac{1}{q^N}\right)$$

and

$$V(q+1, 3p+2) = \frac{q+\frac{2}{3}}{3\sqrt{3}} - \frac{1}{3\pi} + \frac{1}{\pi}\left(q + \frac{2}{3}\right)\left(\log\frac{6\pi}{5} - \gamma\right)$$

$$+ \frac{1}{2}\sum_{k=1}^{\lfloor\frac{N}{2}\rfloor}(-1)^k\frac{4^k\pi^{2k-1}B_{2k}^2}{k(2k)!}\left(\frac{3}{3q+2}\right)^{2k-1}$$

$$+ O\left(\frac{1}{q^N}\right).$$

Based on the relationship [13],

$$V\left(\frac{p}{q}\right) = \frac{1}{\pi}\sum_{r=1}^{q-1}\sum_{k\geq 0}\frac{r(q-2r\bar{p})}{(q(k+1)-r\bar{p})(qk+r\bar{p})},$$

which can be written under the form

$$V(p,q) = \frac{1}{\pi}\sum_{r=1}^{q-1}\sum_{k\geq 0}r\left(\frac{1}{qk+r\bar{p}} - \frac{1}{q(k+1)-r\bar{p}}\right),$$

we transform $V(p/q)$ to the following integral representation [9]:

$$V(p/q) = \frac{1}{\pi} \int_0^1 \frac{P_{q,p}(t)}{(1-t^q)(1-t^p)^2} dt, \tag{53}$$

where $P_{q,p}(t)$ is the rational function

$$P_{q,p}(t) = (q-1)t^{qp+p-1} - qt^{qp-1} + t^{p-1} - t^{p+q-1}$$
$$+ \frac{qt^{2p+q-1} - (q-1)t^{p+q-1}}{t^{pq}}.$$

For $p \equiv 1\,(q)$, we have $\bar{p} = 1$ and $P_{q,1}(t) = (q-2)t^q - qt^{q-1} + qt - q + 2$, and the identity $(5,5)$ (see Ref. [13, Proposition 5.2]) is immediate. In order to evaluate the series expansion of the Vasyunin cotangent sum V, we must investigate the generating function $f(t) = 1/(1-t^q)(1-t^{\bar{p}})^2$ and compute the sequence u_k such that $f(t) = \sum_{k \geq 0} u_k t^k$, and we attract attention that $u_k = b_q(k)$ if $p \equiv 1(q)$. Let k be a positive integer, the double Euclidean algorithm of k over the pair (q,p) is defined [9] by $k = a_k q + b_k p + r_k$ with the conditions that $0 \leq b_k p + r_k < q$ and $0 \leq r_k < p$. The Euclidean algorithm of k over q is written under the form $k = \lfloor k/q \rfloor q + r_k$; similarly, we obtain

$$k = \lfloor k/q \rfloor q + \lfloor (k - \lfloor k/q \rfloor q)/p \rfloor p + r_k.$$

Then $a_k = \lfloor k/q \rfloor$ and $b_k = \lfloor (k - \lfloor k/q \rfloor q)/p \rfloor$. We have shown in Ref. [9, Proposition 4] that

$$u_k = \begin{cases} \left(\left\lfloor \dfrac{a_k}{\bar{p}} \right\rfloor + 1 \right)\left(b_k + 1 + \dfrac{q}{2} \left\lfloor \dfrac{a_k}{\bar{p}} \right\rfloor \right) & \text{if } r_k = 0, \\ 0 & \text{otherwise.} \end{cases}$$

If $r_k = 0$, we say that k is a multiple of (q,p) and the series expansion of $V(p/q)$ follows

$$V(p/q) = \frac{q}{\pi} \sum_{k \equiv 0(q,\bar{p})} \left(\left\lfloor \frac{a_k}{\bar{p}} \right\rfloor + 1 \right)$$

$$\times \left(b_k + 1 + \frac{q}{2} \left\lfloor \frac{a_k}{p} \right\rfloor \right) M_{q,\bar{p}}(k),$$

with

$$N_{q,\bar{p}}(k)M_{q,\bar{p}}(k) = \alpha(q,\bar{p})\,(k+\bar{p})\,(k+\bar{p}+q)$$
$$- \beta(q,\bar{p})\,(k+q-\bar{p}q+2\bar{p})\,(k+\bar{p}q)$$

and

$$N_{q,\bar{p}}(k) = (k+\bar{p}-\bar{p}q+q)\,(k+q-\bar{p}q+2\bar{p})\,(k+\bar{p}q)$$
$$\times\,(k+\bar{p})\,(k+\bar{p}+q)\,(k+\bar{p}q+\bar{p}).$$

5.4. *Generalized Vasyunin cotangent sum and digamma function*

The digamma function is a special case of polygamma function [17] ψ_m defined by

$$\psi_m\,(z) = \frac{d^{m+1}}{dz^{m+1}}\log\Gamma\,(z)\,, \tag{54}$$

where Γ is the gamma function. We have $\psi_0 = \psi$ and another reformulation of V_a as a function of ψ is given by the following theorem.

Theorem 6. *For any integer $a \geq 0$, we get the following result:*

$$V_a\left(\frac{p}{q}\right) = -\frac{\left(1-(-1)^{a+1}\right)q^a}{\pi\,(a+1)}\sum_{k=1}^{q-1}B_{a+1}\left(\left\{\frac{kp}{q}\right\}\right)\psi\left(\frac{k}{q}\right). \tag{55}$$

Explicitly, we have $V_{2a+1}\left(\frac{p}{q}\right) = 0$ and

$$V_{2a}\left(\frac{p}{q}\right) = -\frac{2q^{2a}}{\pi\,(2a+1)}\sum_{k=1}^{q-1}B_{2a+1}\left(\left\{\frac{kp}{q}\right\}\right)\psi\left(\frac{k}{q}\right). \tag{56}$$

In addition, if $a > 1$, taking $\theta_a\,(z) = \psi_a\,(1-z) - \psi_a\,(z)$, one obtains

$$V_{-a}\left(\frac{p}{q}\right) = -\frac{(-1)^a}{\pi q^a\,(a-1)!}\sum_{k=1}^{q-1}\theta_{a-1}\left(\left\{\frac{kp}{q}\right\}\right)\psi\left(\frac{k}{q}\right). \tag{57}$$

Proof. For $a \geq 0$, we have

$$V_a\left(\frac{p}{q}\right) = \frac{q^a}{a+1} \sum_{k=1}^{q-1} B_{a+1}\left(\left\{\frac{kp}{q}\right\}\right) \cot\left(\frac{\pi k}{q}\right)$$

$$= \frac{q^a}{\pi(a+1)} \sum_{k=1}^{q-1} B_{a+1}\left(\left\{\frac{kp}{q}\right\}\right) \left(\psi\left(1-\frac{k}{q}\right) - \psi\left(\frac{k}{q}\right)\right)$$

$$= \frac{q^a}{\pi(a+1)} \sum_{k=1}^{q-1} B_{a+1}\left(\left\{\frac{-kp}{q}\right\}\right) \psi\left(\frac{k}{q}\right)$$

$$- \frac{q^a}{\pi(a+1)} \sum_{k=1}^{q-1} B_{a+1}\left(\left\{\frac{kp}{q}\right\}\right) \psi\left(\frac{k}{q}\right).$$

Therefore, we deduce that

$$V_a\left(\frac{p}{q}\right) = \frac{q^a}{\pi(a+1)} \sum_{k=1}^{q-1} \left(B_{a+1}\left(\left\{\frac{-kp}{q}\right\}\right)\right.$$

$$\left. - B_{a+1}\left(\left\{\frac{kp}{q}\right\}\right)\right) \psi\left(\frac{k}{q}\right)$$

$$= \frac{\left((-1)^{a+1}-1\right)q^a}{\pi(a+1)} \sum_{k=1}^{q-1} B_{a+1}\left(\left\{\frac{kp}{q}\right\}\right) \psi\left(\frac{k}{q}\right),$$

and the relation (56) follows. But for $a > 1$,

$$V_{-a}\left(\frac{p}{q}\right) = -\frac{(-1)^a}{q^a(a-1)!} \sum_{k=1}^{q-1} \psi_{a-1}\left(\left\{\frac{kp}{q}\right\}\right) \cot\left(\frac{\pi k}{q}\right)$$

$$= -\frac{(-1)^a}{q^a(a-1)!\pi} \sum_{k=1}^{q-1} \psi_{a-1}\left(\left\{\frac{kp}{q}\right\}\right)$$

$$\times \left(\psi\left(\frac{q-k}{q}\right) - \psi\left(\frac{k}{q}\right)\right)$$

$$= -\frac{(-1)^a}{q^a\,(a-1)!\pi} \sum_{k=1}^{q-1} \psi_{a-1}\left(\left\{\frac{kp}{q}\right\}\right) \psi\left(\frac{q-k}{q}\right)$$

$$+ \frac{(-1)^a}{q^a\,(a-1)!\pi} \sum_{k=1}^{q-1} \psi_{a-1}\left(\left\{\frac{kp}{q}\right\}\right) \psi\left(\frac{k}{q}\right).$$

Then,

$$V_{-a}\left(\frac{p}{q}\right) = \frac{(-1)^a}{q^a\,(a-1)!\pi} \sum_{k=1}^{q-1} \psi_{a-1}\left(\left\{\frac{kp}{q}\right\}\right) \psi\left(\frac{k}{q}\right)$$

$$- \frac{(-1)^a}{q^a\,(a-1)!\pi} \sum_{k=1}^{q-1} \psi_{a-1}\left(1 - \left\{\frac{kp}{q}\right\}\right) \psi\left(\frac{k}{q}\right)$$

$$= -\frac{(-1)^a}{\pi q^a\,(a-1)!} \sum_{k=1}^{q-1} \left(\psi_{a-1}\left(1 - \left\{\frac{kp}{q}\right\}\right)\right.$$

$$\left. - \psi_{a-1}\left(\left\{\frac{kp}{q}\right\}\right)\right) \psi\left(\frac{k}{q}\right)$$

$$= -\frac{(-1)^a}{\pi q^a\,(a-1)!} \sum_{k=1}^{q-1} \theta_{a-1}\left(\left\{\frac{kp}{q}\right\}\right) \psi\left(\frac{k}{q}\right).$$

$\qquad\qquad\qquad\qquad\qquad\qquad\qquad\qquad\qquad\qquad\square$

For $a = 0$, we get the identity [5]

$$V\left(\frac{p}{q}\right) = -\frac{2}{\pi} \sum_{k=1}^{q-1} B_1\left(\frac{kp}{q}\right) \psi\left(\frac{k}{q}\right).$$

But to better calculate the sum $V_{2a}\left(\frac{p}{q}\right) + V_{2a}\left(\frac{q}{p}\right)$, we give the following theorem.

Theorem 7. *For any positive integer a, we have the following formulas:*

$$V_{2a}\left(\frac{p}{q}\right) = -\frac{2q^{2a}}{\pi\,(2a+1)\,p} \sum_{k=1}^{pq-1} B_{2a+1}\left(\left\{\frac{kp}{q}\right\}\right) \psi\left(\frac{k}{pq}\right). \qquad (58)$$

Proof.   We start by evaluating the sum

$$
\sum_{\substack{k=1 \\ k \not\equiv 0 \ (\mathrm{mod}\ q)}}^{pq-1} B_{2a+1}\left(\left\{\frac{kp}{q}\right\}\right)\psi\left(\frac{k}{pq}\right)
$$

$$
= \sum_{r=1}^{q-1}\sum_{i=0}^{p-1} B_{2a+1}\left(\left\{\frac{rp}{q}\right\}\right)\psi\left(\frac{iq+r}{pq}\right)
$$

$$
= \sum_{r=1}^{q-1} B_{2a+1}\left(\left\{\frac{rp}{q}\right\}\right)\sum_{i=0}^{p-1}\psi\left(\frac{iq+r}{pq}\right)
$$

$$
= \sum_{r=1}^{q-1} B_{2a+1}\left(\left\{\frac{rp}{q}\right\}\right)\left(p\psi\left(\frac{r}{q}\right) - p\psi\left(p\right) + \sum_{i=0}^{p}\psi\left(\frac{p+i}{p}\right)\right).
$$

But we have

$$
\sum_{r=1}^{q-1}\left(p\psi\left(p\right) - \sum_{i=0}^{p}\psi\left(\frac{p+i}{p}\right)\right) B_{2a+1}\left(\left\{\frac{rp}{q}\right\}\right)
$$

$$
= \left(p\psi\left(p\right) - \sum_{i=0}^{p}\psi\left(\frac{p+i}{p}\right)\right)\sum_{r=1}^{q-1} B_{2a+1}\left(\left\{\frac{rp}{q}\right\}\right) = 0.
$$

This implies that

$$
\sum_{r=1}^{q-1} B_{2a+1}\left(\left\{\frac{rp}{q}\right\}\right)\psi\left(\frac{r}{q}\right) = \frac{1}{p}\sum_{k=1}^{pq-1} B_{2a+1}\left(\left\{\frac{kp}{q}\right\}\right)\psi\left(\frac{k}{pq}\right).
$$

$\square$

6.   Symmetric Sum $S_a\left(\frac{p}{q}\right)$

For any positive integer a, we consider the symmetric sum

$$
S_a\left(\frac{p}{q}\right) = V_a\left(\frac{p}{q}\right) + V_a\left(\frac{q}{p}\right),
$$

Bayad–Goubi [7] proved that $S_{2a+1} = 0$,

$$S_{2a}\left(\frac{p}{q}\right) = -\frac{2}{\pi pq \, (2a+1)} \sum_{k=1}^{pq-1} \left(q^{2a+1} B_{2a+1}\left(\left\{\frac{kp}{q}\right\}\right) \right.$$

$$\left. + p^{2a+1} B_{2a+1}\left(\left\{\frac{kq}{p}\right\}\right) \right) \psi\left(\frac{k}{pq}\right) \quad (59)$$

and

$$S_{-2a-1}\left(\frac{p}{q}\right) = \frac{\pi^{2a}}{pq \, (2a)!} \sum_{\substack{k=1 \\ k \not\equiv 0 \pmod{q}}}^{pq-1} \left(\frac{1}{q^{2a}} \cot^{(2a)}\left(\frac{\pi kp}{q}\right) \right.$$

$$\left. + \frac{1}{p^{2a}} \cot^{(2a)}\left(\frac{\pi kq}{p}\right) \right) \psi\left(\frac{k}{pq}\right). \quad (60)$$

For $a \neq 0$,

$$S_{-2a}\left(\frac{p}{q}\right) = -\frac{1}{\pi pq \, (2a-1)!} \sum_{k=1}^{pq-1} \left(\frac{1}{q^{2a-1}} \theta_{2a-1}\left(\left\{\frac{kp}{q}\right\}\right) \right.$$

$$\left. + \frac{1}{p^{2a-1}} \theta_{2a-1}\left(\left\{\frac{kq}{p}\right\}\right) \right) \psi\left(\frac{k}{pq}\right), \quad (61)$$

where

$$\theta_{2a-1}(z) = -(2a-1)!\left(z^{-2a} - 2\sum_{r=0}^{a} \binom{2a}{2r+1} \Theta_{2a} \right.$$

$$\left. \times (2r+1, z) \, z^{2(a-r)-1} \right).$$

and

$$\Theta_a(r, z) = \sum_{m \geq 1} \frac{m^r}{(m^2 - z^2)^a}.$$

Consequently, we have

$$S\left(\frac{p}{q}\right) = -\frac{2}{\pi pq} \sum_{k=1}^{pq-1} \left(qB_1\left(\left\{\frac{kp}{q}\right\}\right) + pB_1\left(\left\{\frac{kq}{p}\right\}\right) \right) \psi\left(\frac{k}{pq}\right).$$

$$(62)$$

Thereafter,

$$S\left(\frac{p}{q}\right) = -\frac{1}{\pi pq} \sum_{r=\min\{p,q\}}^{pq-1} \left(2pB_1\left(\left\{\frac{r}{p}\right\}\right) - 2qB_1\left(\left\{\frac{r}{q}\right\}\right)\right.$$

$$\left. - \sigma_{p,q}(r)\right)\sigma_{p,q}(r)\psi\left(\frac{r}{pq}\right) \quad (63)$$

and

$$S\left(\frac{p}{q}\right) = -\frac{1}{\pi pq} \sum_{r=\min\{p,q\}}^{pq-1} \left(2qB_1\left(\left\{\frac{r}{q}\right\}\right) - 2pB_1\left(\left\{\frac{r}{p}\right\}\right)\right.$$

$$\left. - \sigma_{p,q}(-r)\right)\sigma_{p,q}(-r)\psi\left(1 - \frac{r}{pq}\right), \quad (64)$$

where

$$\sigma_{p,q}(r) = p\sigma_p(r) - q\sigma_q(r)$$

and

$$\sigma_p(r) = \left\{\frac{r}{p}\right\} - \left\{\frac{r-1}{p}\right\}.$$

6.1. *Series expansion of the symmetric sums $S(\frac{p}{q})$*

From the integral representations of $V(p/q)$ and $V(q/p)$, we obtain

$$S\left(\frac{p}{q}\right) = \frac{1}{\pi}\int_0^1 (Q_{q,\bar{p}}(t) + Q_{p,\bar{q}}(t))\,dt.$$

Thanks to symmetric function G and hint of the relation [7]

$$\pi S(p,q) = G(p,p) + G(q,q) - 2G(p,q)$$

$$+ (q-p)\log\frac{p}{q}, \quad (65)$$

we give others integral and series expansion of $S(p,q)$. For simplifying calculus, we consider the auxillary function $\theta_{p,q}$ given by

$$\theta_{p,q}(r) = \left(p\left\{\frac{r}{p}\right\} - q\left\{\frac{r}{q}\right\}\right)$$

$$\times \left(p\left\{\frac{r}{p}\right\} - q\left\{\frac{r}{q}\right\} + q - p\right). \quad (66)$$

Let $\dot{r}$ and $\ddot{r}$ be, respectively, the remainder of the Euclidean division of r over p and q. Then we get $\theta_{p,q}(r) = (\dot{r} - \ddot{r})(\dot{r} - \ddot{r} + q - p)$. Substituting the expression (66) and the value $\log \frac{p}{q} = p\frac{\log p}{p} - q\frac{\log q}{q}$ in the identity (65), we deduce that

$$\pi S(p, q) = \sum_{k \geq 1} \frac{\theta_{p,q}(k)}{k(k+1)}.$$

We have already proved the following theorem.

Theorem 8. *The series expansion of $S(p, q)$ is given by the following equivalent identities:*

$$S\left(\frac{p}{q}\right) = \frac{1}{\pi} \sum_{k \geq 1} \frac{\left(\dot{k} - \ddot{k}\right)\left(\dot{k} - \ddot{k} + q - p\right)}{k(k+1)} \tag{67}$$

and

$$S\left(\frac{p}{q}\right) = \frac{1}{\pi} \sum_{k \equiv 0(pq)} \sum_{r=p}^{pq-1} \frac{(\dot{r} - \ddot{r})(\dot{r} - \ddot{r} + q - p)}{(k+r)(k+r+1)}. \tag{68}$$

Consequently, the integral representation of $S(p, q)$ is given by the following corollary.

Corollary 2.

$$S\left(\frac{p}{q}\right) = \frac{1}{\pi} \int_0^1 \frac{\sum_{r=p}^{pq-1} (\dot{r} - \ddot{r})(\dot{r} - \ddot{r} + q - p)(1 - t) t^{r-1}}{1 - t^{pq}} dt. \tag{69}$$

Proof. Since we have

$$\frac{1}{(k+r)(k+r+1)} = \int_0^1 \left(t^{k+r-1} - t^{k+r}\right) dt.$$

Let $\sum = \sum_{k \equiv 0(pq)} \frac{1}{(k+r)(k+r+1)}$, then

$$\sum = \int_0^1 \left(t^{r-1} - t^r\right) \sum_{i \geq 0} t^{ipq} dt$$

$$= \int_0^1 \frac{\left(t^{r-1} - t^r\right)}{1 - t^{pq}} dt.$$

According to identity (68), Theorem 8, the result (69) holds true. $\square$

Under the expression of $S\left(\frac{p}{q}\right)$, another reformulation of the Vasyunin formula is given by the following expressions:

$$\langle e_p, e_q \rangle = \frac{\log 2\pi - \gamma}{2} \left(\frac{1}{p} + \frac{1}{q}\right) + \frac{p-q}{2pq} \log \frac{q}{p}$$

$$- \frac{1}{2pq} \sum_{k \equiv 0(pq)} \sum_{r=p}^{pq-1} \frac{(\dot{r} - \ddot{r})(\dot{r} - \ddot{r} + q - p)}{(k+r)(k+r+1)}$$

and

$$\langle e_p, e_q \rangle = \frac{\log 2\pi - \gamma}{2} \left(\frac{1}{p} + \frac{1}{q}\right) + \frac{p-q}{2pq} \log \frac{q}{p}$$

$$- \frac{1}{2pq} \sum_{k \geq 1} \frac{\left(\dot{k} - \ddot{k}\right)\left(\dot{k} - \ddot{k} + q - p\right)}{k(k+1)}.$$

Acknowledgments

We are grateful to M.Th. Rassias who proposed me to write this chapter.

References

[1] M. Abramowitz and I. A. Stegun (1972). *Handbook of Mathematical Function with Formula, Graphs and Mathematical Tables* (tenth printing with corrections, United States Department of Commerce, New York).

[2] T. M. Apostol (1976). *Introduction to Analytic Number Theory* (Springer-Verlag, New York, Heidelberg, Berlin).

[3] L. Báez-Duarte, M. Balazard, M. Landreau, and E. Saias (2005). Etude de l'autocorrélation multiplicative de la fonction partie fractionnaire. *Ramanujan J.*, **9**, 215–240.

[4] L. Báez-Duarte (2003). A strengthening of the Nyman–Beurling criterion for the Riemann hypothesis. *Alti Accad. Naz. Lincei Cl. Sci. Fis. Mat. Natur. Rend. Lincei (9) Mat. Appl.*, **14**(1), 5–11.

[5] L. Báez-Duarte, M. Balazard, B. Landreau, and E. Saias, (2000). Notes sur la fonction ζ de Riemann, 3. *Adv. Math.*, **149**, 130–144.

[6] M. Balazard and A. de Roton (2010). Sur un critère de Báez-Duarte pour l'hypothèse de Riemann. *Int. J. Number Theory*, **6**(4), 883–903.

[7] A. Bayad and M. Goubi (2018). Reciprocity formulae for generalized Dedekind–Vasyunin-cotangent sums. *Math. Methods Appl. Sci.*, **42**(4), 1082–1098. https://doi.org/10.1002/mma.5414.

[8] A. Bayad and M. Goubi (2013). Proof of the Möbius conjecture revisited. *Proc. Jangjeon Math. Soc.*, **16**(2), 237–243.

[9] S. Belhadj and M. Goubi (2020). On the Vasyunin cotangent sums related to Riemann hypothesis. *WSEAS Trans. Math.*, **19**, 676–682.

[10] S. Belhadj and M. Goubi (2022). On Bell polynomials associated to Vasyunin cotangent sums. *Int. Electron. J. Algebra*, **31**(31), 230–242.

[11] S. Bettin and J. B. Conrey (2013). Period functions and cotangent sums. *Algebra Number Theory*, **7**(1), 215–242.

[12] S. Bettin and J. B. Conrey (2013). A reciprocity formula for a cotangent sum. *Int. Math. Res. Not.*, (24), 5709–5726.

[13] M. Goubi, A. Bayad, and M. O. Hernane (2017). Explicit and asymptotic formulae for Vasyunin-cotangent sums. *Publ. Inst. Math. (N. S.)*, **102**(116), 155–174.

[14] M. Goubi (2020). On the cotangent sums related to Estermann zeta function and arithmetic properties of their arguments *WSEAS Trans. Math.*, **19**(6), 57–64.

[15] M. Goubi (2021). Series expansion of a cotangent sum related to the Estermann zeta function, *Kragujevac J. Math.*, **45**(3), 343–352.

[16] J. Katz and Y. Lindell (2014). *Introduction to Modern Cryptography*, Second Edition (CRC Press, Taylor & Francis Group, Boca Raton, London, New York).

[17] K. S. Kölbig (1996). The polygamma function and the derivatives of the cotangent function for rational arguments. CERN-IT-Reports, CERN-CN-96-005.

[18] H. Maier and M. T. Rassias (2015). The order of magnitude for moments for certain cotangent sums. *J. Math. Anal. Appl.*, **429**(1), 576–590.

[19] H. Maier and M. T. Rassias (2016). Generalizations of a cotangent sum associated to the Estermann zeta function. *Commun. Contemp. Math.*, **18**(1), 1550078.

[20] H. Maier and M. T. Rassias (2017). The maximum of cotangent sums related to Estermann zeta function in rational numbers in short intervals. *Appl. Anal. Discrete Math.*, **11**, 166–176.

[21] H. Maier and M. T. Rassias (2020). Cotangent sums related to the Riemann hypothesis for various shifts of the argument, *Canad. Math. Bull.*, **63**(3), 522–535.

M. Goubi

[22] I. Niven and H. S. Zuckerman (1972). *An Introduction to the Theory of Numbers* Third Edition (Wiley Eastern Limited, New York).

[23] B. Nyman (1950). On Some Groups and Semigroups of Translations, Thesis, Uppsala.

[24] M. T. Rassias (2014). A cotangent sum related to zeros of the Estermann zeta function. *Appl. Math. Comput.*, **240**, 161–167.

[25] V. I. Vasyunin (1995). On a biorthogonal system associated with the Riemann hypothesis. *Algebra i Analiz*, **7**(3), 118–135.

Chapter 7

The Approximation of Inverse Functions of Dirichlet Series by Rational Functions

Johann Franke

*Universität zu Köln, Department Mathematik Gyrhofstraße
8b 50931 Köln, Germany*

jfrank12@uni-koeln.de

Extending the classical work of Kalmár, we give a method to construct inverse functions for convergent Dirichlet series with nonnegative coefficients on positive intervals in their region of convergence.

1. Introduction

Let $s = \sigma + it \in \mathbb{C}$ and

$$D(s) := \sum_{n=1}^{\infty} a_n n^{-s}$$

be a Dirichlet series with nonnegative coefficients a_n such that $a_n = O(n^A)$ for some $A > 0$. It is easy to see that $D(s)$ converges absolutely for values s with $\sigma > A + 1$. It then represents a holomorphic function on the half plane $\{s \in \mathbb{C} \mid \sigma > A + 1\}$. According to a theorem by Landau, there is a unique number $\sigma_a \in \mathbb{R} \cup \{-\infty\}$, such that $D(s)$ converges (absolutely) for $\sigma > \sigma_a$ and diverges for $\sigma < \sigma_a$. When restricting to the real axis, one sees that the function $x \mapsto D(x)$ is monotonously decreasing in the region of convergence and can therefore be inverted on intervals $I \subset (\sigma_a, \infty)$. Put $D(\sigma_a) := \lim_{\sigma \to \sigma_a^+} D(\sigma)$. In particular, assuming $a_1 = 0$ for a while, there is an inverse function $D^{-1} : (0, D(\sigma_a)) \to (\sigma_a, \infty)$. In these notes, we provide a method to give approximations for the inverse D^{-1} in terms of rational functions in the coefficients a_n of $D(s)$. To state the main result, we define for each integer $n > 0$ a polynomial $P_n(X_2, \ldots, X_n, Y)$ with n unknowns by

$$P_n(X_2, \ldots, X_n, Y) := 1 + \sum_{\ell \geq 1} \sum_{\substack{j_1, \ldots, j_\ell > 1 \\ j_1 \cdots j_\ell \leq n}} X_{j_1} \cdots X_{j_\ell} Y^\ell.$$

A few explicit examples for the first P_n are given in the following.

Example 1. We have

$$P_2(X_2; Y) = 1 + X_2 Y$$

$$P_5(X_2, X_3, X_4, X_5; Y) = 1 + (X_2 + X_3 + X_4 + X_5)Y + X_2^2 Y^2$$

$$P_8(X_2, \ldots, X_8; Y) = 1 + \left(\sum_{j=2}^{8} X_j \right) Y$$

$$+ (X_2^2 + 2X_2 X_3 + 2X_2 X_4)Y^2 + X_2^3 Y^3$$

Define also corresponding rational functions Q_n in $2n$ unknowns by

$$Q_n(X_2, \ldots, X_{2n}, Y) := \frac{P_{2n}(X_2, \ldots, X_{2n}, Y)}{P_n(X_2, \ldots, X_n, Y)}.$$

Then our main theorem is the following.

Theorem 1. Let $D(s) = \sum_{n \geq 2} a_n n^{-s}$ be a Dirichlet series that converges somewhere, such that $a_n \geq 0$ and there exist integers m, k with $a_m \neq 0 \neq a_k$ and $m^a \neq k^b$ for all integers $a, b > 0$. Then we have for *all $0 < \lambda < D(\sigma_a)$*

$$\lim_{n \to \infty} Q_n \left(a_1, \ldots, a_{2n-1}, \frac{1}{\lambda} \right) = \exp(\rho_\lambda \log(2)),$$

where $\rho_\lambda := D^{-1}(\lambda)$ is the unique solution of the equation $D(x) = \lambda$ in the interval (σ_a, ∞).

Remark. In the case that a_n are strictly multiplicative, the expressions $Q_n \left(a_1, \ldots, a_{2n-1}, \frac{1}{\lambda} \right)$ take a much simpler form. More details of this case are discussed in the following.

Dirichlet series are of great importance in analytic number theory. Already Euler noticed that if the coefficients a_n satisfy the strictly multiplicative relation $a_n a_m = a_{nm}$, one can expand the series in the range of absolute convergence into a product ranging over all primes:

$$\sum_{n=1}^{\infty} a_n n^{-s} = \prod_p \frac{1}{1 - a_p p^{-s}}.$$

This beautiful observation has many far-reaching consequences. For instance, when putting $a_n = 1$, one obtains the famous *Riemann zeta function*

$$\zeta(s) := \sum_{n=1}^{\infty} \frac{1}{n^s}, \qquad (\sigma > 1),$$

which has been studied intensively over the last 150 years. Since the coefficients of $\zeta(s)$ are clearly strictly multiplicative, one has

$$\zeta(s) = \prod_p \frac{1}{1 - p^{-s}}, \qquad (\sigma > 1).$$

The Euler product for the Riemann zeta function is just the starting point for a series of celebrated theorems regarding prime number distribution. After Bernhard Riemann's discoveries of the connection between the prime numbers and the zeros of the globally meromorphic continuation of $\zeta(s)$, in 1896, Hadamard and de la Vallée Poussin proved the prime number theorem, which states that the number

$\pi(x)$ of primes smaller than x satisfies the asymptotic equivalence

$$\pi(x) \sim \int_2^x \frac{dt}{\log(t)} =: \mathrm{Li}(x), \qquad x \to \infty.$$

Assuming the famous *Riemann hypothesis* (RH), stating that all zeros ρ of $\zeta(s)$ satisfy $\mathrm{Re}(\rho) \le \frac{1}{2}$, the prime number theorem can be improved to

$$\pi(x) = \mathrm{Li}(x) + O(\sqrt{x}\log(x)), \qquad x \to \infty. \tag{1}$$

A proof that (1) implies the RH, and vice versa, can be found in Chapter II of [13]. In addition to numerous applications in many areas of mathematics, RH is also of interest in cryptography. For example, the RSA cryptosystem uses large prime numbers to construct a public as well as a private key. Its security is based on the fact that there is still no efficient algorithm for conventional computers to decompose a number into its prime factors. The theory behind RSA requires only results from elementary number theory. For a detailed introduction, we refer to Koblitz's book [8], especially Chapter IV. Again based on simple number theory, using Fermat's Little Theorem, Miller [10] developed a deterministic primality test, in 1976, that works under the assumption of the so-called *Generalized Riemann Hypothesis* (GRH). GRH states that all nontrivial zeros of the Dirichlet L-functions $L(s, \chi)$ have real part $\frac{1}{2}$, and like RH, it remains unproved until today. In 1980, Rabin [11] used Miller's results to develop a probabilistic test that works independent of GRH. Through the work by Bach [1] in 1990, this so-called Miller–Rabin test can be transformed into a deterministic test running with speed $O((\log n)^4)$, again assuming GRH. This, as well as many other applications in number theory, contributed to the inclusion of the Riemann conjecture on the list of the seven Millennium Problems by the Clay Mathematics Institute in 2000. The interested reader is advised to consult the great survey articles by Bombieri [2] and Sarnak [12] for more background information.

2. Ordered Factorization of Natural Numbers and a Theorem by Kalmár

Following a notation by Hille [5], for a fixed positive integer k, let

$$f_k(n) := \#\{(n_1, \ldots, n_k) \in \mathbb{N}_{>1}^k \mid n_1 \cdots n_k = n\}. \tag{2}$$

We then define $f(1) := 1$ and for $n > 1$,

$$f(n) = \sum_{k=1}^{\infty} f_k(n),$$

i.e. $f(n)$ is the number of possibilities to write a number n as a product of positive integers greater than 1 such that the order of factors matters. For example, one has $f(12) = 8$, since

$$12 = 12$$
$$= 2 \cdot 6 = 6 \cdot 2$$
$$= 3 \cdot 4 = 4 \cdot 3$$
$$= 2 \cdot 2 \cdot 3 = 2 \cdot 3 \cdot 2 = 3 \cdot 2 \cdot 2.$$

In Ref. [6], Kalmár showed 1931 the following remarkable theorem.

Theorem 2 (Kalmár). *Let $\rho > 1$ be the real number satisfying $\zeta(\rho) = 2$, where $\zeta(s)$ is the Riemann zeta function. Then we have*

$$\sum_{n \leq x} f(n) \sim \frac{x^\rho}{\rho |\zeta'(\rho)|}, \qquad x \to \infty.$$

We have $\rho = 1.7286472389\ldots$ and $\frac{1}{\rho |\zeta'(\rho)|} = 0.3181736521\ldots$. Kalmár's work was refined by Warlimont [14] by considering only products whose factors satisfy additional conditions. Hille [5] showed that for arbitrarily small $\delta > 0$, the inequality $f(n) > n^{\rho-\delta}$ holds for infinitely many n, i.e.

$$\limsup_{n \to \infty} \frac{f(n)}{n^{\rho-\delta}} = \infty.$$

Several refinements of Hille's work for specific families of n where given by Chor, Lemke and Mador [3]. MacMahon [9] provided a closed formula for $f(n)$ in terms of the prime factor decomposition of n (see (5)), but in the context of "multipartite numbers". Apart from that, the function $f(n)$ does not seem to have received a lot of attention.

As far as the author knows, no closed representation for the constant ρ in Theorem 2 in terms of elementary constants is known. For more background information on ρ, and other composition constants, the reader may wish to consult [4].

The key result for a quick proof of Theorem 2 is the following Taberian Theorem (see [13] on p. 350 for a more general version).

Theorem 3 (Wiener–Ikehara, Delange). *Let* $D(s) := \sum_{n=1}^{\infty} a_n n^{-s}$ *be a Dirichlet series with* $a_n \geq 0$, *converging for* $\mathrm{Re}(s) > \sigma > 0$, *such that* $s \mapsto D(s) - \frac{A}{s-\sigma}$ *has a continuous continuation to* $\{s \in \mathbb{C} \mid \mathrm{Re}(s) \geq \sigma\}$. *Then one has*

$$\sum_{0<n\leq x} a_n \sim \frac{A}{\sigma} x^{\sigma}, \qquad x \to \infty.$$

We sketch the proof of Theorem 2. Define $D(s) = \sum_{n=1}^{\infty} \frac{f(n)}{n^s}$. We first note that

$$D(s) = \sum_{p=0}^{\infty} \left(\sum_{k=2}^{\infty} \frac{1}{k^s} \right)^p = \sum_{p=0}^{\infty} (\zeta(s) - 1)^p = \frac{1}{2 - \zeta(s)},$$

for $|\zeta(s) - 1| < 1$, which means that the function D has a meromorphic continuation to the entire plane and a simple pole at $s = \rho > 1$, whereas $\zeta(\rho) = 2$ with residue

$$\mathrm{res}_{s=\rho} D(s) = -\frac{1}{\zeta'(\rho)} = \frac{1}{|\zeta'(\rho)|}.$$

An easy argument proves $\zeta(\rho + it) \neq 2$ für all $t \neq 0$. One can now apply Theorem 3 to $D(s)$ to obtain Theorem 2.

The purpose of this chapter is to generalize Kalmár's Theorem to a statement about inverse functions of Dirichlet series with nonnegative coefficients. For the construction of the inverse functions, Theorem 3 will mainly be used.

3. Construction of Inverse Functions of Dirichlet Series

Consider the subset of all formal Dirichlet series $D(s) := \sum_{n=1}^{\infty} a_n n^{-s}$ satisfying the following properties:

(i) We have $a_1 = 0$ and $a_n \geq 0$.
(ii) There are integers $m, \ell > 1$, such that $m^a \neq \ell^b$ for all $a, b \in \mathbb{N}$, with $a_m \neq 0 \neq a_\ell$.
(iii) The series $D(s)$ converges somewhere, i.e. we have $\sigma_a < \infty$ for the absolute abscissa of convergence.

Remember that we put $D(\sigma_a) := \lim_{\sigma \to \sigma_a^+} D(\sigma) \in \mathbb{R}^+ \cup \{\infty\}$, where σ_a is the abscissa of convergence of D. We want to call this subset of Dirichlet series $\mathcal{D}^*$. Note that $\mathcal{D}^* \cup \{0\}$ has a monoid structure via addition. Also, we have $\lambda \mathcal{D}^* = \mathcal{D}^*$ for all $\lambda > 0$. Note that condition (ii) is equivalent to $\{n \in \mathbb{N} \mid a_n \neq 0\} \not\subset \{k, k^2, k^3, \ldots\}$ for all $k > 1$, since all integers m, ℓ with $m^a = \ell^b$ are powers of some common integer k and vice versa. For the abscissa of convergence $-\infty \leq \sigma_a < \infty$, we find by monoticity that $D|_{(\sigma_a, \infty)} : (\sigma_a, \infty) \to (0, D(\sigma_a))$ has an inverse function $D^{-1} : (0, D(\sigma_a)) \to (\sigma_a, \infty)$. For $\lambda > 0$, we define $\sigma_a < \rho_\lambda$ to be the (unique) number satisfying $D(\rho_\lambda) = \lambda$.

The following key lemma also shows that $\mathcal{D}^*$ is precisely the right class of Dirichlet series to look at.

Lemma 1. *Let $D(s) = \sum_{n=2}^{\infty} a_n n^{-s}$ be a nonconstant Dirichlet series with real nonnegative coefficients and abscissa of convergence $\sigma_a < \infty$. Then we have the equivalences:*

(i) $D \in \mathcal{D}^*$.
(ii) *For all $0 < \lambda < D(\sigma_a)$, we have $D(\rho_\lambda + it) \neq \lambda$ for all $t \neq 0$.*

Proof. Firstly, we show that $D(\rho_\lambda + it) = \lambda$ for some $t \neq 0$ implies $D \notin \mathcal{D}^*$. We have $D(\rho_\lambda + it) = \lambda$ if and only if

$$\sum_{n=2}^{\infty} a_n \cos(t \log(n)) n^{-\rho_\lambda} = \lambda \tag{3}$$

and

$$\sum_{n=2}^{\infty} a_n \sin(t \log(n)) n^{-\rho_\lambda} = 0 \tag{4}$$

are both satisfied. Since $a_n \cos(t\log(n))n^{-\rho_\lambda} \leq a_n n^{-\rho_\lambda}$, equation (3) is satisfied if and only if $a_n \cos(t\log(n))n^{-\rho_\lambda} = a_n n^{-\rho_\lambda}$. Hence, we have $\cos(t\log(n)) = 1$ always if $a_n \neq 0$. As a result, we obtain $t\log(n) = 2\pi m$ for some m. For values $n_1 \neq n_2$ such that $a_{n_1} \neq 0 \neq a_{n_2}$ we find as $t \neq 0$, the relation $\frac{1}{m_1}\log(n_1) = \frac{1}{m_2}\log(n_2)$ for some positive integers $m_1 \neq m_2$. Hence, $n_1^{m_2} = n_2^{m_1}$. We conclude that $\{n \in \mathbb{N} \mid a_n \neq 0\} \subset \{k, k^2, k^3, \ldots\}$ for some $k > 1$. But this implies $D \notin \mathcal{D}^*$.

On the other hand, we easily see that $D \notin \mathcal{D}^*$ implies (3) and (4) for $t = \frac{2\pi}{\log(k)} > 0$, and as a result, $D(\rho_\lambda + it) = \lambda$. $\qquad\square$

Next, for $n > 1$, we introduce the notation

$$U_n(X_2, \ldots, X_n; Y) := \sum_{k=1}^{\infty} Y^k \sum_{\substack{j_1, \ldots, j_k > 1 \\ j_1 \cdots j_k = n}} X_{j_1} \cdots X_{j_k}$$

and also put $U_1(Y) := 1$. Examples are

$$U_2(X_2; Y) = X_2 Y, \qquad U_3(X_2, X_3; Y) = X_3 Y,$$
$$U_4(X_2, X_3, X_4; Y) = X_2^2 Y^2 + X_4 Y.$$

Note that

$$P_n(X_2, \ldots, X_n; Y) = \sum_{k=1}^{n} U_k(X_2, \ldots, X_k, Y).$$

We are now ready for the proof of our main theorem.

Proof of Theorem 1. Let $0 < \lambda < D(\sigma_a)$. Consider the function

$$\widehat{D}_\lambda(s) := \frac{1}{1 - \frac{D(s)}{\lambda}}.$$

Since we have $|D(s)| < \lambda$ for all $s \in \mathbb{C}$ with $\mathrm{Re}(s) > \rho_\lambda$, the function $\widehat{D}_\lambda(s)$ is holomorphic on the half plane $\{s \in \mathbb{C} \mid \mathrm{Re}(s) > \rho_\lambda > 0\}$.

In this region, we find a Dirichlet series representation

$$\frac{1}{1 - \frac{D(s)}{\lambda}} = 1 + \sum_{k=1}^{\infty} \left(\frac{D(s)}{\lambda} \right)^k$$

$$= 1 + \sum_{k=1}^{\infty} \lambda^{-k} \sum_{n=2}^{\infty} \left(\sum_{\substack{j_1,\ldots,j_k>1 \\ j_1 \cdots j_k = n}} a_{j_1} \cdots a_{j_k} \right) n^{-s}$$

$$= 1 + \sum_{n=2}^{\infty} U_n(a_2, \ldots, a_n; \lambda^{-1}) n^{-s}.$$

By Lemma 1, we conclude that the function

$$s \mapsto \widehat{D}(s) - \frac{\lambda}{|D'(\rho_\lambda)|(s - \rho_\lambda)}$$

has a holomorphic continuation to $\{s \in \mathbb{C} \mid \mathrm{Re}(s) \geq \rho_\lambda\}$. As the coefficients are all nonnegative, by the Tauberian Theorem 3, we conclude

$$\sum_{0<k\leq n} U_k(a_2, \ldots, a_k; \lambda^{-1}) = P_n(a_2, \ldots, a_n, \lambda^{-1}) \sim \frac{\lambda n^{\rho_\lambda}}{\rho_\lambda |D'(\rho_\lambda)|}.$$

It follows

$$\lim_{n\to\infty} \frac{P_{2n}(a_2, \ldots, a_{2n}, \lambda^{-1})}{P_n(a_2, \ldots, a_n, \lambda^{-1})} = 2^{\rho_\lambda},$$

and this completes the proof. □

4. The Case of Strictly Multiplicative Coefficients

4.1. *General theory*

The main result can be further refined in the case that the coefficients a_n of the Dirichlet series are strictly multiplicative. In this case, we find

$$P_n(a_2, \ldots, a_n; \lambda) = 1 + \sum_{\ell=2}^{n} U_k(a_2, \ldots, a_\ell; \lambda)$$

$$= 1 + \sum_{\ell=2}^{n} \sum_{m=1}^{\infty} \lambda^m \sum_{\substack{j_1, \ldots, j_m > 1 \\ j_1 \cdots j_m = \ell}} a_{j_1} \cdots a_{j_m}$$

$$= 1 + \sum_{\ell=2}^{n} a_k \sum_{m=1}^{\infty} \lambda^m \sum_{\substack{j_1, \ldots, j_m > 1 \\ j_1 \cdots j_m = \ell}} 1$$

$$= 1 + \sum_{\ell=2}^{n} a_\ell \sum_{m=1}^{\infty} \lambda^m f_m(\ell),$$

where $f_m(\ell)$ was defined in (2). Note that the polynomials $g_\ell(\lambda) := \sum_{m=1}^{\infty} \lambda^m f_m(\ell)$ do not depend on the coefficients $a_2, a_3, a_4, \ldots$ of the Dirichlet series. Examples are

$$g_2(\lambda) = \lambda, \quad g_{10}(\lambda) = \lambda + 2\lambda^2, \quad g_{16}(\lambda) = \lambda + 3\lambda^2 + 3\lambda^3 + \lambda^4,$$

$$g_{360}(\lambda) = \lambda + 22\lambda^2 + 111\lambda^3 + 220\lambda^4 + 190\lambda^5 + 60\lambda^6.$$

Knowing the decomposition of ℓ into prime factors, one can easily compute the numbers $f_m(\ell)$ and with this the terms $g_\ell(\lambda)$. For the convenience of the reader, we provide a short argumentation. Let

$$P_k(n) := \#\{(n_1, \ldots, n_k) \in \mathbb{N}^k \mid n_1 \cdots n_k = n\}.$$

Note that we obtain

$$\zeta(s)^k = \sum_{n=1}^{\infty} P_k(n) n^{-s}.$$

In particular, $P_k(n)$ is multiplicative for all k. In the following proposition, we provide a formula for $P_k(n)$ in terms of the prime factorization of n.

Proposition 2. *Let* $n = p_1^{\nu_1} p_2^{\nu_2} p_3^{\nu_3} \cdots p_\ell^{\nu_\ell}$ *the prime factorization of* n. *For all* $k \in \mathbb{N}$, *we then have*

$$P_k(n) = \prod_{j=1}^{\ell} \left| \binom{-k}{\nu_j} \right|.$$

Proof. Let $n = n_1 \cdots n_k$ for integers $n_1, \ldots, n_k \in \mathbb{N}$. Then we have

$$n_j = p_1^{\nu_{1,j}} \cdots p_\ell^{\nu_{\ell,j}} \qquad \text{with} \quad \nu_i = \sum_{j=1}^{k} \nu_{i,j}.$$

Hence, there is a one-to-one correspondence between tuples $(n_1, \ldots, n_k)$ as above and matrices

$$\begin{pmatrix} \nu_{1,1} & \cdots & \nu_{\ell,1} \\ \nu_{1,2} & \cdots & \nu_{\ell,2} \\ \vdots & \vdots & \vdots \\ \nu_{1,k} & \cdots & \nu_{\ell,k} \end{pmatrix} \in \mathbb{N}_0^{k \times \ell}$$

with the property

$$\sum_{j=1}^{k} (\nu_{1,j}, \nu_{2,j}, \ldots, \nu_{\ell,j}) = (\nu_1, \ldots, \nu_\ell).$$

Now, there are $\left| \binom{-k}{\nu_i} \right|$ ways to write ν_i as the sum of k nonnegative integers such that the order of summands matters, hence there are

$$\prod_{j=1}^{\ell} \left| \binom{-k}{\nu_j} \right|$$

of such matrices which proves the proposition. $\qquad \square$

There is the following relationship between the functions f_k and P_k. We note the generating function

$$(\zeta(s) - 1)^k = \sum_{n=1}^{\infty} f_k(n) n^{-s}.$$

The following proposition shows us how to compute the values $f_k(n)$ when knowing the prime decomposition of n.

Proposition 3. *We have*

$$f_k(n) = \sum_{j=0}^{k} \binom{k}{j} (-1)^{k-j} P_j(n).$$

Proof. With

$$(\zeta(s) - 1)^k = \sum_{j=0}^{k} \binom{k}{j}(-1)^{k-j}\zeta(s)^j,$$

we obtain by comparing coefficients

$$f_k(n) = \sum_{j=0}^{k} \binom{k}{j}(-1)^{k-j}P_j(n).$$

This proves the proposition.   □

Note that Knopfmacher and Mays [7] provide the explicit formula

$$f(p_1^{\nu_1}p_2^{\nu_2}\cdots p_k^{\nu_k}) = \sum_{j=1}^{\nu_1+\cdots+\nu_k}\sum_{m=0}^{j-1}(-1)^m\binom{j}{m}\prod_{h=1}^{k}\binom{\nu_h + j - m - 1}{\nu_h},$$

$$\tag{5}$$

which seems to have been first discovered by MacMahon in 1893. Note that we have the identity $\left|\binom{-a}{b}\right| = \binom{a+b-1}{b}$ for integers $a, b > 0$.

4.2. *Application to the zeta function*

We apply our results to the case of the Riemann zeta function. We find

$$\sum_{0 < m \leq 50000} g_m(\lambda) = 50000\lambda + 448726\lambda^2 + 1851017\lambda^3$$

$$+ 4637675\lambda^4 + 7859697\lambda^5 + 9497358\lambda^6$$

$$+ 8404336\lambda^7 + 5508642\lambda^8 + 2675372\lambda^9$$

$$+ 953575\lambda^{10} + 244267\lambda^{11} + 43374\lambda^{12}$$

$$+ 4954\lambda^{13} + 330\lambda^{14} + 16\lambda^{15} =: L_{50000}(\lambda),$$

and

$$\sum_{0<m\leq 25000} g_m(\lambda) = 25000\lambda + 207037\lambda^2 + 783349\lambda^3$$
$$+ 1786119\lambda^4 + 2727466\lambda^5 + 2933053\lambda^6$$
$$+ 2274989\lambda^7 + 1282772\lambda^8 + 523286\lambda^9$$
$$+ 151748\lambda^{10} + 30262\lambda^{11} + 3872\lambda^{12}$$
$$+ 287\lambda^{13} + 15\lambda^{14} =: L_{25000}(\lambda).$$

Now, we find with Theorem 1 that

$$\omega_{50000}(x) := \frac{1}{\log(2)} \log\left(\frac{L_{50000}(\frac{1}{x})}{L_{25000}(\frac{1}{x})}\right)$$

approximates an inverse of $s \mapsto \zeta(s) - 1$ in the region $0 < x < \infty$. Figure 1 shows a plot of the function $x \mapsto \mathrm{Re}(\omega_{50000}(x))$. Note that $\mathrm{Re}(\omega_{50000}(x)) = \omega_{50000}(x) > 0$ for $x > 0$. When composing the function $y \mapsto \zeta(y) - 1$ with $y = \omega_{50000}(x)$, we obtain Figure 2. We observe that the approximation also seems to hold in a (real-valued) neighborhood of the singularity of $\zeta(s) - 1$ in in $s = 1$. This suggests to us that the method shown works in much larger domains than the positive real axis. Indeed, Figures 3 and 4 gives numerical evidence that it extends to large parts of the complex number plane.

5. Open Questions

For a Dirichlet series with certain properties, we can approximate the associated inverse function in the domain of the positive real axis by modified sequences of rational functions. This procedure seems to extend in some cases to larger domains in the complex plane. There remain, therefore, some questions whose investigation may be of independent interest.

(1) Given a Dirichlet series $D(s)$ satisfying all required assumptions, how large is the domain of convergence of the associated sequence of rational functions?

(2) Let $\sigma_a > 0$. Is there a (simple) criterion for deciding, on the basis of the coefficients $a_1, a_2, \ldots$ of $D(s)$, whether it has a locally

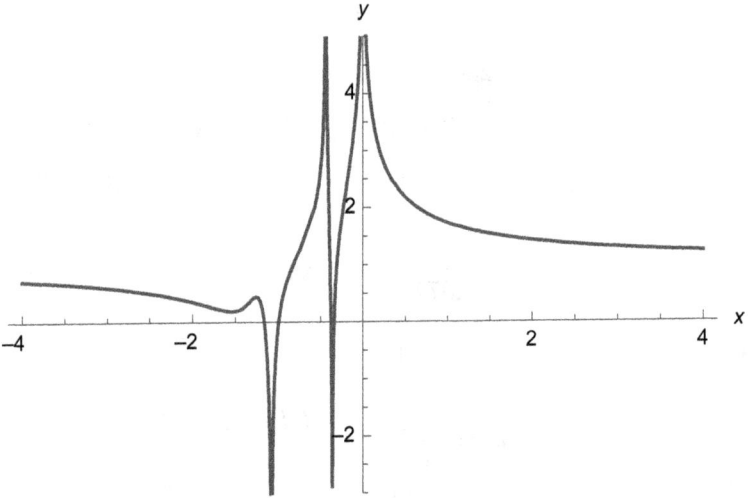

Figure 1.   The graph of $\mathrm{Re}(\omega_{50000}(x))$.

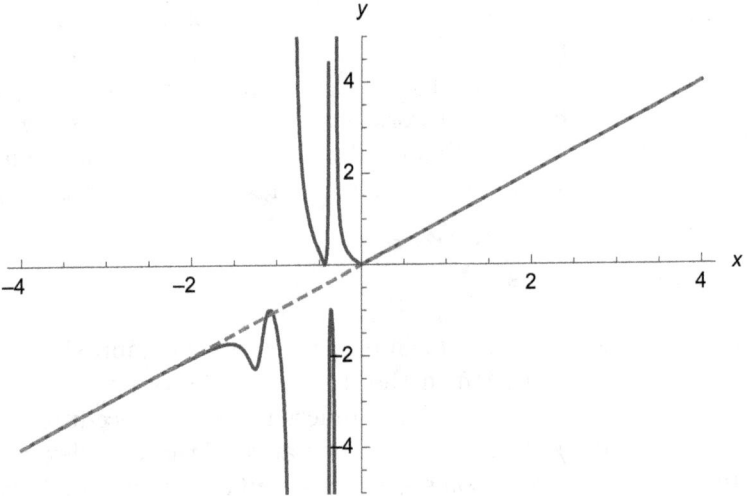

Figure 2.   The graph of $\zeta(\mathrm{Re}(\omega_{50000}(x))) - 1$. The identity $x \mapsto x$ is shown as a dashed graph.

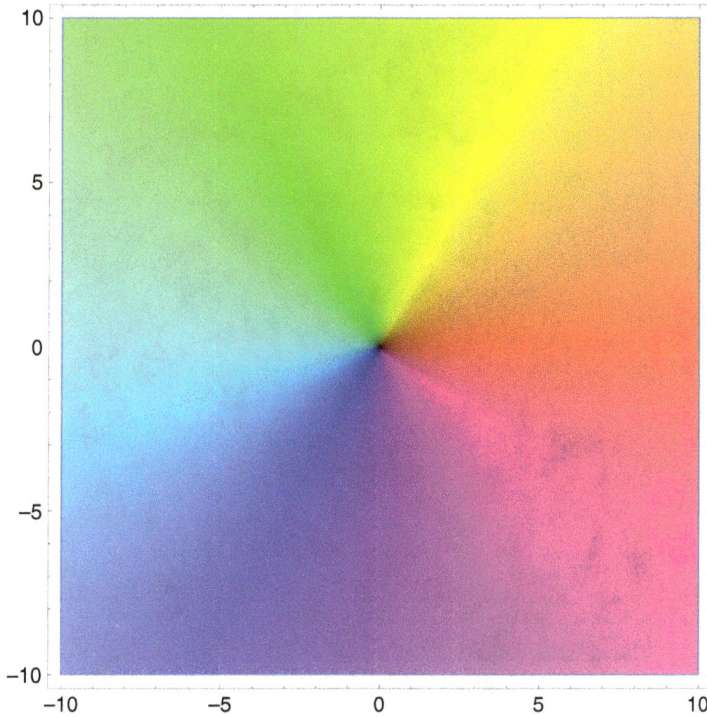

Figure 3. Complex graph of the identity function.

meromorphic continuation around its boundary singularity $s = \sigma_a$, that is, whether it can be expanded into a Laurent series of the form

$$D(s) = \frac{b_{-1}}{s - \sigma_a} + b_0 + \cdots ,$$

where b_ν are numbers depending on $a_1, a_2, \ldots$? Since D is supposed to have an inverse around its singularity, the degree of the pole must be 1.

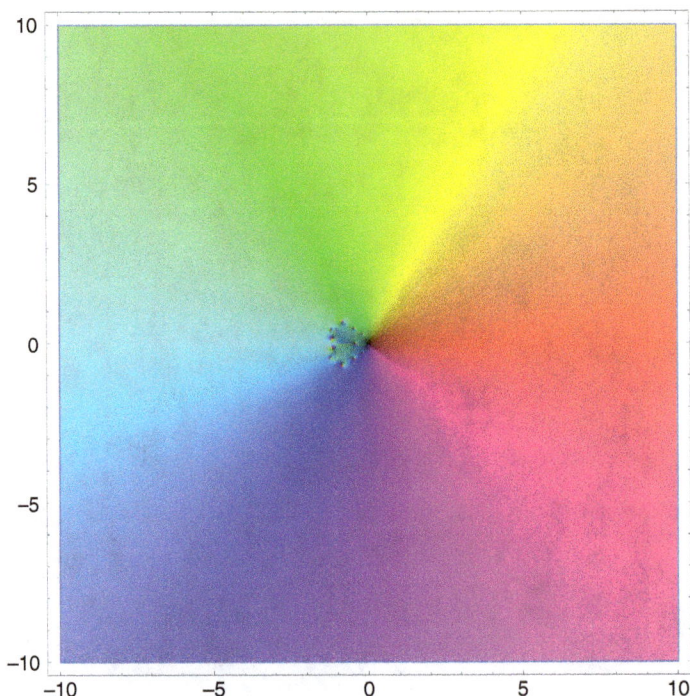

Figure 4.　The function $z \mapsto \zeta(\mathrm{Re}(\omega_{50000}(z))) - 1$.

Note that a theorem in the spirit of (2) has in a certain sense the character of an abelian theorem. The generating coefficients of the Dirichlet series under consideration are to be used to infer its local behavior at the boundary of the region of convergence.

References

[1] E. Bach (1990). Explicit bounds for primality testing and related problems. *Math. Comput.*, **55**(191), 355–380.
[2] E. Bombieri (2000). Problems of the Millennium: The Riemann Hypothesis. Official Problem Description, https://www.claymath.org/sites/default/files/official_problem_description.pdf.
[3] B. Chor, P. Lemke, and Z. Mador (2000). On the number of ordered factorizations of natural numbers. *Disc. Math.*, **214**, 123–133.
[4] S. R. Finch (2003). *Mathematical Constants* (Cambridge University Press, New York, 2003).

[5] E. Hille (1936). A problem in 'factorisatio numerorum.' *Acta Arith.*, **2**, 134–144.

[6] L. Kalmár (1931). A factorisatio numerorum problémájáról, *Mat. Fiz. Lapok*, **38**, 1–15.

[7] A. Knopfmacher and M. Mays (2006). Ordered and unordered factorizations of integers. *Mathematica J.*, **10**, 72–89.

[8] N. Koblitz (1994). *A Course in Number Theory and Cryptography*, Second Edition (Springer-Verlag, New York).

[9] P. A. MacMahon (1893). Memoir on the theory of the compositions of numbers. *Philos. Trans. R. Soc. London A*, **184**, 835–901.

[10] G. L. Miller (1976). Riemann's hypothesis and tests for primality. *J. Comput. Syst. Sci.*, **13**(3), 300–317.

[11] M. O. Rabin (1980). Probabilistic algorithm for testing primality. *J. Number Theory*, **12**(1), 128–138.

[12] P. Sarnak (2004). *Problems of the Millennium: The Riemann Hypothesis*. CMI Annual Report, 5–21.

[13] G. Tenenbaum (1995). Introduction to analytic and probabilistic number theory. *Am. Math. Soc.*, **163**.

[14] R. Warlimont (1993). Factorisatio numerorum with constraints. *J. Number Theory*, **45**(2), 186–199.

Chapter 8

Topological Data Analysis and Clustering

Dimitrios Panagopoulos

Eureka Module, Pelopa 3, Gerakas,
Attiki, Greece
Department of Mathematics, National University of Athens,
Athens, Greece

dpanagop@gmail.com
https://www.linkedin.com/in/dpanagopoulos/

Clustering is one of the most common tasks of machine learning. In this chapter, we examine how ideas from topology can be used to improve clustering techniques.

1. Introduction

With the advent of Big Data, algorithms that try to extract information from them are ubiquitous. Clustering algorithms are a subcategory of machine learning algorithms with a wide range of applications. Notions like closeness, distance and shape are central to clustering. It is then natural to try to use ideas and techniques from topology to improve clustering algorithms. This chapter examines some ways on how this could be achieved. In Section 2, a brief introduction to the clustering task is presented. In Section 2.1, a definition is presented along with notation. k-means algorithm will be one of our focus areas and is examined in more detail in Section 2.2. In Section 3, topological data analysis (TDA) is presented. We will be using persistent homology and the Mapper algorithm which are presented in Sections 3.1 and 3.2, respectively. Ways of applying topology to clustering are examined in Section 4. Finally, Section 5 has some concluding remarks.

All figures in this chapter were created using Scikit-TDA package [26] in Python. The relative code can be found in https://github .com/dpanagop/data_analytics_examples/tree/master/topological_ data_analysis.

2.   Clustering

2.1.   *Definition*

Clustering is the task of grouping a set of objects in such a way that objects in the same group (called a cluster) are more similar (in some sense) to each other than to those in other groups [32]. Clustering is a major task of machine learning with diverse applications from medicine [29] to sports [1].

The definition is intentionally general and very common. When one tries to apply clustering algorithms, a lot of thought is to be devoted to how similarity should be defined. For example, in marketing, a common application is customer clustering. There one should consider if similarity depends on things like age, gender, place of residence, education, etc. It must be clear that in most cases, there is no predetermined grouping of the objects to be clustered, against which an algorithm can be measured.[1]

In order to have some notation available, we state the following problem.

Problem 1 (Clustering Problem). *Given a finite subset $X \subset \mathbb{R}^d$, find a partition $C_1, \ldots, C_k$ of X such that elements in every subset C_i are more similar to other elements in C_i than to elements in another subset C_j, where $j \neq i$.*

---

[1]In Machine Learning, the category of problems where there is no given labeling or values that can be used to train a model is called unsupervised learning.

Some of the most common algorithms used for clustering are as follows:

(i) k-means,
(ii) hierarchical clustering,
(iii) density-based spatial clustering of applications with noise (DBSCAN),
(iv) Gaussian mixture models.

In the following section, k-means will be described in more detail. For the rest, the interested reader can start by reading the related Wikipedia article [32], the documentation of Python's Scikit-learn package [24] at https://scikit-learn.org/stable/modules/clustering. html and [30].

2.2. k-Means

One of the most popular algorithms for clustering is k-means. It is an algorithm that finds an approximate solution to the following problem.

Problem 2 (k-Means Problem). *Given a finite subset $X \subset \mathbb{R}^d$ and a natural number $k > 1$, find a partition $C_1, \ldots, C_k$ of X that minimizes*

$$\sum_{i=1,\ldots,k} \sum_{x \in C_i} ||x - c_i||^2, \tag{1}$$

where $c_i = \frac{1}{|C_i|} \sum_{x \in C_i} x$ is the barycenter of C_i.

While k-means problem is known to be NP-hard [2,20], the k-means algorithm is an efficient heuristic that detects a local minimum. The algorithm starts with a random selection of k centers $c_1, \ldots, c_k$. Then

(1) each $x \in X$ is assigned to its closest center, thus

$$C_i = \{x : ||x - c_i|| < ||x - c_j||, \, j \in \{1, \ldots, k\} \setminus \{i\}\}, \quad i = 1, \ldots, k,$$

(2) new centers are calculated based on equation $c_i = \frac{1}{|C_i|} \sum_{x \in C_i} x, \, i = 1, \ldots, k,$
(3) steps 1 and 2 are repeated until centers $c_1, \ldots, c_k$ do not change or a maximum number of iterations is reached.

There are several variations to avoid reaching a local minima and/or helping the algorithm converge faster. Those variations consist mainly of changing the way the initial centers are selected (e.g. see Refs. [4,23]). Furthermore, there are extensions for dealing with datasets X with great number of points [27] or for replacing centers with real data points [16].

k-means is a very efficient algorithm that converges fast and has, relatively, low memory requirements when implemented in a computer system. Two drawbacks of k-means are as follows:

- the number of clusters (value of k) cannot be determined by the algorithm, instead it is an input for the algorithm,
- k-means can detect clusters that have a spherical shape.[2]

For example, in Figure 1, a natural way to cluster the points would be in two concentric rings. k-means cannot achieve this. It is also

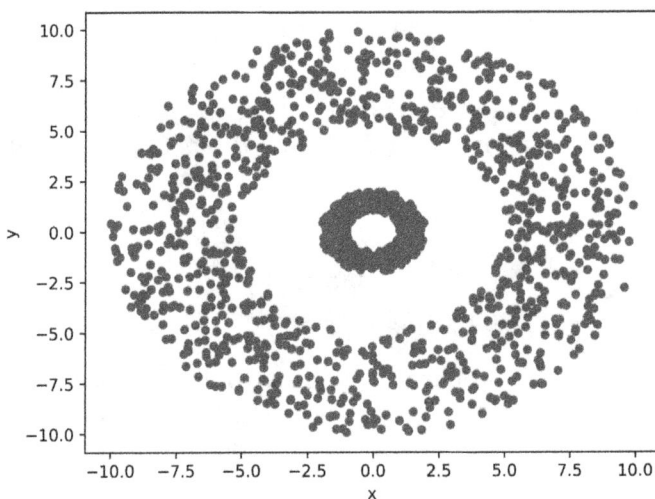

Figure 1.   The image shows 1500 points in the plane. The points were selected at random such that 500 lie in a ring with inner radius 1 and outer 2. The remaining 1000 were selected such that they lie in a ring with inner radius 5 and outer 10.

---

[2]It can be shown that there is no algorithm that is unaffected from scaling and from the act of packing more densely elements of the same cluster that can detect all possible partitions (i.e. that can detect clusters of arbitrary shapes) [18].

worth mentioning that k-means is not the only clustering algorithm that requires as input the number of clusters.

3. Topological Data Analysis

Assuming that X is a sample from a manifold, the number clusters is the number of connected components of the manifold. Furthermore, the shape of the connected components is related to the topological features of the manifold. Thus, it is natural to turn to the mathematical field of topology for ideas to deal with the above problems. This gives rise to the field of TDA. In this section, we present two of its tools: topological persistence and the Mapper algorithm.

3.1. *Persistent homology*

Topological Persistence was introduced by Edelsbrunner, Letscher and Zomorodian in 2002 [10]. Let $X = \{x_1, \ldots, x_n\} \subset \mathbb{R}^d$ be a finite set of points. One can use X to create a simplicial complex as follows.

Definition 1 (Vietoris–Rips Complex). Let $X = \{x_1, \ldots, x_n\} \subset \mathbb{R}^d$ be a finite set of points and $\epsilon > 0$ a positive real number. The Vietoris–Rips complex $\mathcal{C}_{\mathrm{VR}}(X, \epsilon)$ is the complex with X as vertex set and where the n-simplex $\{x_{i_0}, \ldots, x_{i_n}\}$ belongs to $\mathcal{C}_{\mathrm{VR}}(X, \epsilon)$ if and only if $\|x_{i_s} - x_{i_t}\| < \epsilon$ for all $0 \le s, t \le n$.

The Vietoris–Rips complex can be seen as an approximation of the manifold X is sampled from and the parameter ϵ as the maximum distance between two points that are supposed to be path connected.

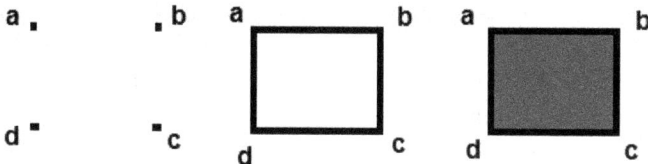

Figure 2. Four points on the corners of a unit square and the corresponding Vietoris–Rips complex for various selections of ϵ. To the left when $\epsilon \in (0, 1)$, in the middle when $\epsilon \in [1, \sqrt{2})$, and to the right when $\epsilon \ge \sqrt{2}$.

A standard construction of algebraic topology, takes a simplicial complex $\mathcal{C}$ and constructs a family of abelian groups and connecting homomorphisms

$$\cdots \to C_{n+1} \xrightarrow{\partial_{n+1}} C_n \xrightarrow{\partial_n} C_{n-1} \to \cdots \to C_1 \xrightarrow{\partial_1} C_0 \xrightarrow{\partial_0} 0$$

with $\partial_n \partial_{n+1} = 0$, i.e. $\mathrm{Im}\partial_{n+1} \subseteq \mathrm{Ker}\partial_n$ and where C_n has as a base the n-simplices of $\mathcal{C}$. Thus, one can define the **nth homology group**

$$H_n := \mathrm{Ker}\partial_n / \mathrm{Im}\partial_{n+1}.$$

It can be shown if a complex has a finite number of n-simplices, then the nth homology group is a finitely generated abelian group. Thus, in this case, the nth homology group is isomorphic to the direct sum of copies of $\mathbb{Z}$ and finite cyclic groups, i.e. $H_n \simeq \mathbb{Z}^b \bigoplus_{i=1,\ldots,m} \mathbb{Z}_{p_i}$. The number of copies $\mathbb{Z}$ appears (i.e. b) is called the **nth Betti number**, β_n. Intuitively, b_0 counts the number of connected components, b_1 the number of one-dimensional holes, b_2 the number of two-dimensional holes, etc. For a more detailed exposition, the reader can see Ref. [15, Chapter 2].

It is clear that if $\epsilon < \epsilon'$, then every n-simplex of $\mathcal{C}_{\mathrm{VR}}(X, \epsilon)$ is also a k-simplex of $\mathcal{C}_{\mathrm{VR}}(X, \epsilon')$, hence $\mathcal{C}_{\mathrm{VR}}(X, \epsilon) \subseteq \mathcal{C}_{\mathrm{VR}}(X, \epsilon')$. It follows that the following diagram where the vertical maps are inclusions is commutative:

$$\mathcal{C}_{\mathrm{VR}}(X, \epsilon) : \cdots \to C_{n+1} \xrightarrow{\partial_{n+1}} C_n \xrightarrow{\partial_n} C_{n-1} \to \cdots,$$
$$\downarrow \qquad \downarrow \qquad \downarrow$$
$$\mathcal{C}_{\mathrm{VR}}(X, \epsilon') : \cdots \to C'_{n+1} \xrightarrow{\partial'_{n+1}} C'_n \xrightarrow{\partial'_n} C'_{n-1} \to \cdots,$$

and hence, the inclusion map defines a homomorphism between nth homology groups $\mathcal{C}_{\mathrm{VR}}(X, \epsilon)$ and nth homology groups of $\mathcal{C}_{\mathrm{VR}}(X, \epsilon')$ (see Ref. [15, Proposition 2.9]). It is then possible for every dimension n to keep track of when (i.e. for what value of ϵ) a new generator is added to the nth homology group and when it vanishes. This can be depicted in a two-dimensional diagram showing where each generator corresponds a point (x, y), where x is the value of ϵ, the generator is created and y the value of ϵ' the generator vanishes. The farther away a point is from the main diagonal, the more the corresponding

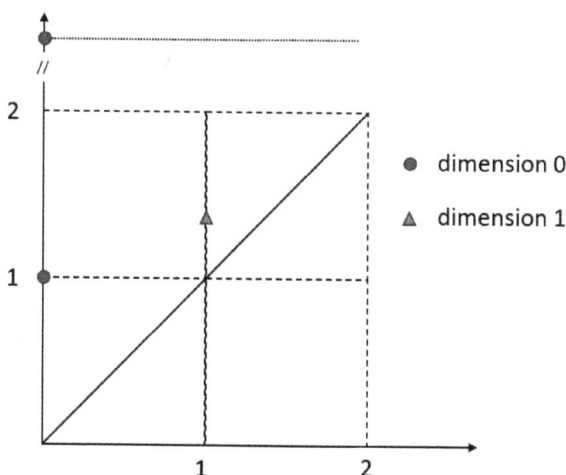

Figure 3.   Persistent diagram for the Vietoris–Rips complexes of Figure 2. There are three points at $(0,1)$ that represent three of the four connected components when $\epsilon < 1$. Since for $\epsilon = 1$, all four vertices are connected, three of the four generators of H_1 vanish. The remaining generator does not vanish. This corresponds to the point that lies in the intersection of x-axis and the horizontal line at the top that represents infinity. For dimension 1, there is a loop that is created when $\epsilon = 1$ and destroyed when $\epsilon = \sqrt{2}$. This corresponds to the point that is marked with a triangle.

generator survives (see Figure 3). The idea behind persistent homology is that topological features that survive longer will correspond to actual features of the data and to random noise. For more details, one can read chapters [6,8,10]. References [9,12] cover persistent homology as well as a more broad view of applications of topology.

Note that there is another way to depict persistence by using the so-called barcode diagrams. In barcode diagrams, each generator is represented by a horizontal bar with starting point of the value of ϵ where the generator is created and end point of the value of ϵ where the generator vanishes. The bars are stacked one above the other with the bar that has the smaller starting point at the top (e.g. see Figure 5(b)).[3]

---

[3]Note that the barcode diagrams presented here does not contain bars ending at infinity.

3.2. *Mapper*

The Mapper algorithm was presented by Singh, Mémoli and Carlsson [28]. The authors state the following:

> The basic idea can be referred to as partial clustering, in that a key step is to apply standard clustering algorithms to subsets of the original dataset, and then to understand the interaction of the partial clusters formed in this way with each other.

The Mapper algorithm starts with a finite subset $X \subset \mathbb{R}^d$, a map $f : X \to Z$ to a topological space Z and a covering $\{U_\alpha\}_{\alpha \in A}$ of Z.[4] Then, the following can be

(1) Every nonempty set $f^{-1}(U_\alpha)$ is clustered using a clustering algorithm. Thus, for every $\alpha \in A$, where $f^{-1}(U_\alpha) \neq \emptyset$, we have a finite set $\{C_{\alpha 1}, \ldots, C_{\alpha k_\alpha}\}$.
(2) A simplicial complex is constructed with vertex set of all clusters $\{C_{\alpha j} : 1 \leq j \leq k_\alpha, f^{-1}(U_\alpha) \neq \emptyset\}$ and where an n-simplex $\{C_{\alpha_0 i_0}, \ldots, C_{\alpha_n i_n}\}$ belongs to the complex if and only if $C_{\alpha_0 i_0} \cap \cdots \cap C_{\alpha_n i_n} \neq \emptyset$.

In the original chapter by Singh, Mémoli and Carlsson, single-linkage clustering [33] was used in step (1). More details on Mapper can be found in Refs. [6,14,28].

4. Applications of TDA to Clustering

4.1. *Using persistent homology*

It is obvious that persistent homology can be used to obtain information about the shape of data. Especially, it can provide information on the number of connected components and help decide whether data have spherical shape or not. In this chapter, we present the following examples:

- **Two squares:** In this example, 100 points were selected at random using the uniform distribution so that they are in a square with vertices $(0,0)$, $(1,0)$, $(1,1)$, $(0,1)$. Similarly, 100 points were

---

[4]Usually, $Z = \mathbb{R}$ or $Z = \mathbb{R}^2$.

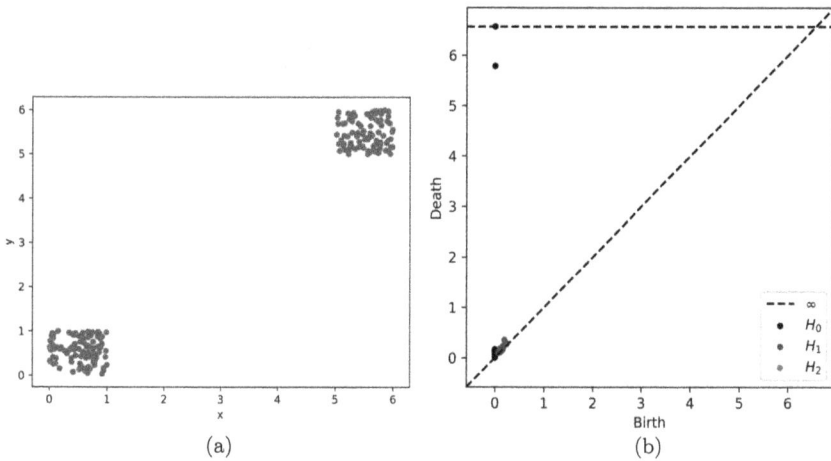

Figure 4. Two squares made up from random points (a) and the corresponding persistent diagram (b).

selected in a square with vertices $(5,5)$, $(6,5)$, $(6,6)$, $(5,6)$. The dataset is depicted in Figure 4.

The corresponding persistent diagram is in Figure 4. Note that in dimension 0, there is a group of points near the origin, a point near $(0,6)$[5] and a point with zero x-coordinate on the horizontal dashed line that represents infinity. This is a clear indication that the corresponding Vietoris–Rips complex $\mathcal{C}(X,\epsilon)$ has many connected components for small values of ϵ. Those can be attributed to random noise. The other two points indicate that, at large scale, there exist two connected components that are merged into one for $\epsilon \simeq 6$.[6] For dimension 1, all the points lie close to the diagonal, hence it can be deduced that the dataset does not contain any holes.

- **Two circles:** In this example, 500 points were selected at random using the uniform distribution so that they are in a ring with inner radius 1 and outer 2. Similarly, 1000 points were selected in a ring with inner radius 5 and outer 10. The dataset is depicted in Figure 1.

---

[5] The coordinates are $(0, t')$, where t' is equal to the closest distance between the two squares which is approximately $4\sqrt{2}$.

[6] Actually, $\epsilon = t'$.

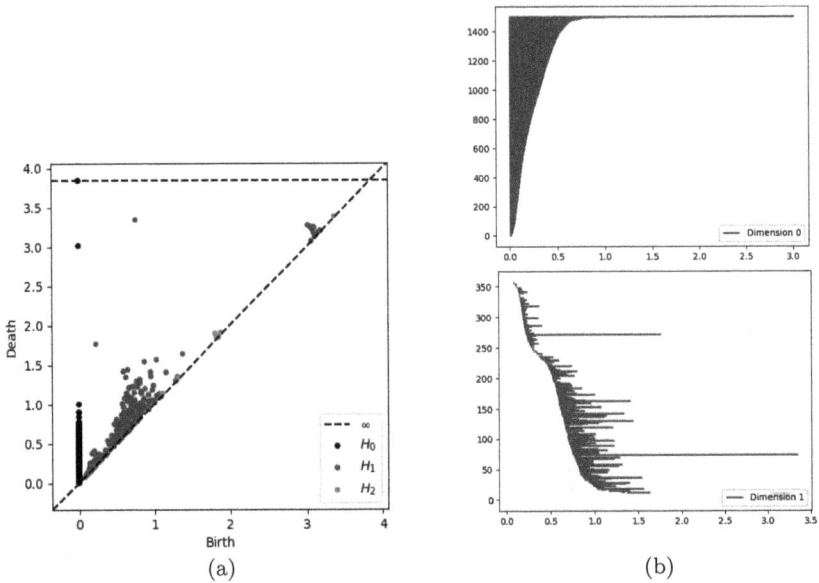

Figure 5.   Persistent diagram and barcode diagrams for the two circles dataset.

The corresponding persistent diagram is in Figure 5. In dimension 0, there is a group of points with zero x-coordinate and y-coordinate ranging from zero to approximately one. There is also a point at a height little before two and one lying at the dashed line representing infinity. As in the case of two squares, this is an indication that there are two distinct groups. In dimension 1, note several points above the diagonal that indicate the existence of holes. Most of them are two points, one at approximately (0.4, 1.8) and one at approximately (0.6, 3.5) (Figure 5) that represent the two circles. This should act as a warning against using k-means.

- **Iris dataset:** The Iris dataset [3,11] contains measurements of petal and sepal length and width from 150 samples from three species (*Setosa, Versicolor, Virginica*) of Iris. There are 50 samples from each of *Iris setosa, Iris versicolor* and *Iris Virginica* species Figure 6.

The corresponding persistent diagram is in Figure 7. In dimension 0, one can clearly see two groups as indicated by the two top points with zero x-coordinate. It is not clear from the diagram if the dataset can be split in one, two or three more groups.

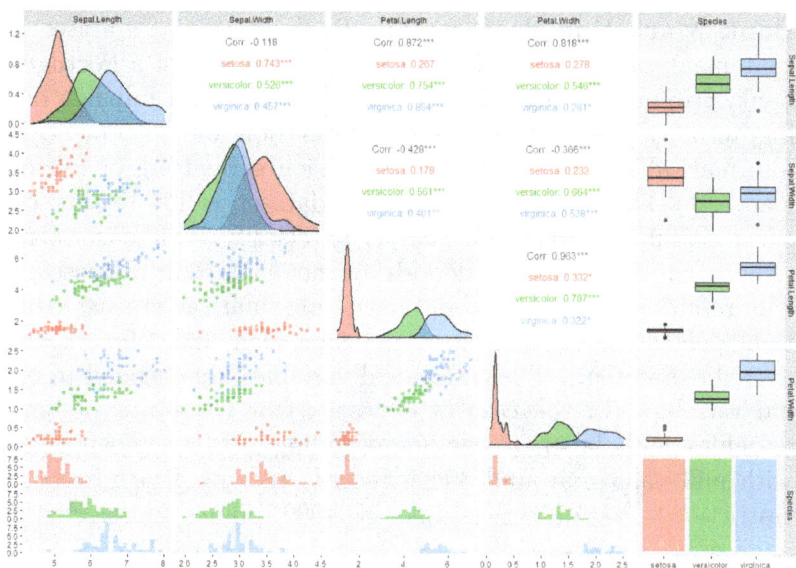

Figure 6. Plot of Iris dataset. In the main diagonal, there are the histograms for each variable for each of the three species. Pairwise scatter plots (resp. correlations for the whole dataset and per specie) are below (resp. above) the main diagonal. The last column to the right contains the box plots for each variable and specie.

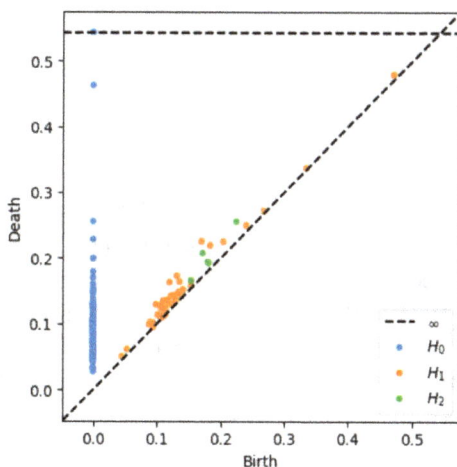

Figure 7. Persistent diagram of the Iris dataset.

The points that correspond to dimensions 1 and 2 lie close to the diagonal, hence there is no indication that k-means will not be an effective clustering method.

- **Bank Marketing dataset:** The Bank Marketing dataset [21] contains data for approximately 45000 customers of a Portuguese bank. Specifically, the data are from a marketing campaign offering term deposits. It contains information on customers age, job, marital status, education, possession of housing or personal loan, etc. The standard use of the dataset is for testing classification algorithms. The goal is to classify each customer as either acquiring a term deposit or not. For our purposes, we will retain information about age, job, marital status, education, possession of housing or personal loan and whether there is credit in default. The categorical variables are encoded to ordinal variables (in the case of education) or to groups of binary variables with label or one-hot encoding. After removing cases with null values, around 30000 records remain. Due to resource restrictions, a random sample of 4000 points is selected for analysis.

From the persistence diagram (Figure 8), it can be seen that there are three or four clusters. Furthermore, it is clear that the dataset contains one-dimensional holes, and thus, it is not prone to clustering by k-means.

4.2. *Using Mapper*

In this section, we present the results of application of the Mapper algorithm to the above datasets. For the two squares and two circles datasets, the projection in the first coordinate was used as map $f : X \to Z$ (see Section 3.2). For the Iris dataset and Bank Marketing datasets projection on principal components derived by PCA was used. It should be noted that since this is an introduction, we will not present an exhaustive analysis of the results.

- **Two squares:** In Figure 9, we can see the result of Mapper on the dataset where the points are selected at random within two squares. Each square is represented by two connected nodes.

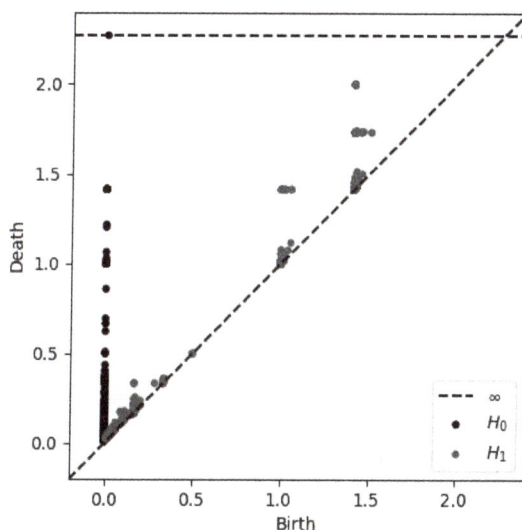

Figure 8. Persistent diagram of the Bank Marketing dataset.

Figure 9. Application of the Mapper algorithm on two squares dataset.

- **Two circles:** While in the case of two squares Mapper represents the two separate clusters of points, this is not the case for the two circles. In Figure 10, we can see the results when using DBSCAN or k-means for clustering. The expected result would be a graph consisting of nodes forming two cycles. This was not possible to achieve even though a variety of parameters and clustering methods were tested. Figure 10 is an indication of the high sensitivity of the Mapper algorithm to the selection of clustering algorithms.
- **Iris dataset:** For the Iris dataset, projection on the first two principal components was used as map $f : X \to Z$ (Section 3.2). It is clear from the pair plots that Setosa, one of the three species of

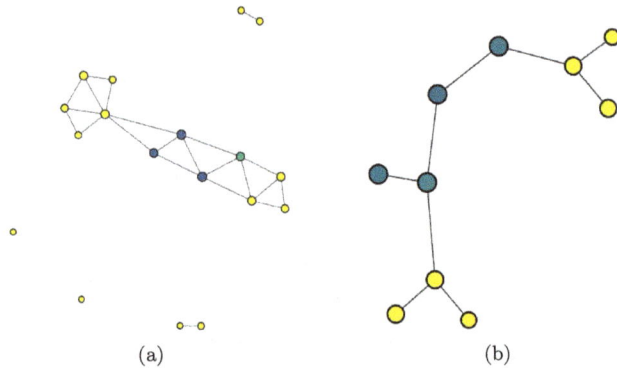

Figure 10.   Application of Mapper to the two circles. Projection to the first coordinate was used. For (a), DBSCAN was used as clustering algorithm, while k-means was used for (b). Color indicates ratio between points from the outer and points from the inner circles in each node, ranging from yellow for nodes where all points are from the outer circle to blue for nodes with all points from the inner circle.

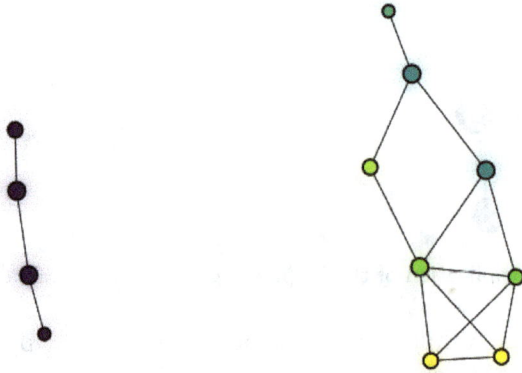

Figure 11.   Application of Mapper to Iris dataset. Color indicates species. Purple for Setosa, green for Versicolor and yellow for Virginica.

Iris, is clearly separated from the other two (see Figure 6). Mapper manages to capture this fact. As we can see, Mapper constructs a graph with two connected components. One of them corresponds to Setosa. The other corresponds to the other two species. Furthermore, in the second connected component, nodes on one side correspond to Versicolor and nodes on the other side correspond to Virginica (Figure 11).

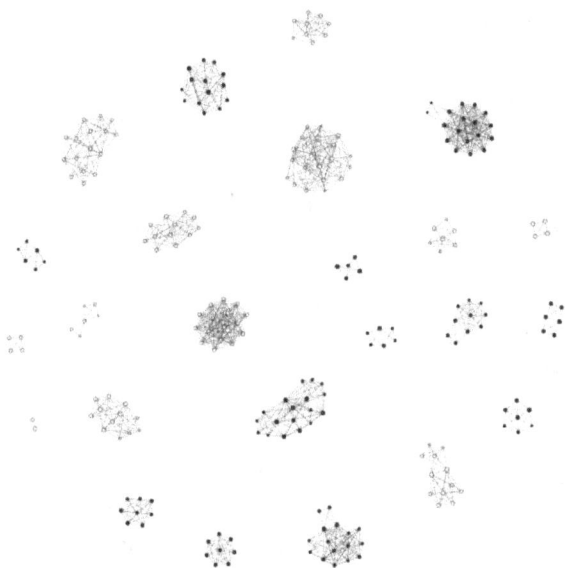

Figure 12. Application of Mapper to Bank Marketing dataset.

- **Bank Marketing dataset:** For Bank Marketing dataset, projection on the first five principal components was used as map $f : X \to Z$ (see Section 3.2). Mapper reveals, several well-defined groups of customers (see Figure 12). By examining the values of the variables, we are able to determine which of them are used to define those groups. This process reveals that the existence of a housing loan or the marital status play a major role. In contrast, age does not.

5. Concluding Remarks

In the previous sections, we applied persistent homology and Mapper to some simple datasets. In particular, the experiments we conducted indicate that persistent homology can be used successfully to study the number of clusters and the shape of a dataset. In contrast, the Mapper algorithm produced mixed results, in some cases managing to capture key information about the data, while in some other cases failing to do so.

More applications can be found in Refs. [6,9,10]. In Ref. [13], the authors use TDA to detect relationships between products sold on a local level and products sold on a national level. It is a very interesting idea that it also highlights one of the obstacles in using TDA: the fact that the current algorithms do not scale to accommodate for Big Data. In Ref. [19], the authors apply Mapper to analyze gene expression in breast tumors, performance data from the NBA and voting data from US House of Representatives. In Ref. [17], the authors compare Mapper against k-means and hierarchical clustering on the task of image popularity in social media.

It is clear from the applications above that persistent homology and the Mapper algorithm can be used to provide insights on datasets. Unfortunately, there are a couple of reasons that make wider adoption of these tools difficult. Some of them are the rather technical background required to understand these tools, the existence of some well-established techniques for addressing similar problems and the lack of high efficient algorithms that scale well to Big Data. That said, the field is active with research going on both on applications and on the theoretical background, (e.g. see Refs. [5,7,25]).

References

[1] S. E. Akhani (2019). *Distance Construction and Clustering of Football Player Performance Data*, PhD Thesis. https://discovery.ucl.ac.uk/id/eprint/10065964/1/thesis.pdf.

[2] D. Aloise, A. Deshpande, P. Hansen, and P. Popat (2009). NP-hardness of Euclidean sum-of-squares clustering. *Mach. Learn.*, **75**(2), 245–248.

[3] E. Anderson (1935). The irises of the Gaspe Peninsula. *Bull. Amer. Iris Soc.*, **59**, 2–5.

[4] D. Arthur and S. Vassilvitskii (2007). k-means++: The advantages of careful seeding. *Proceedings of the Eighteenth Annual ACM-SIAM Symposium on Discrete Algorithms*, (Society for Industrial and Applied Mathematics). http://ilpubs.stanford.edu:8090/778/1/2006-13.pdf.

[5] S. Basu and L. Parida (2017). Spectral sequences, exact couples and persistent homology of filtrations. *Expo. Math.*, **35**(1), 119–132. https://www.sciencedirect.com/science/article/pii/S0723086916300378.

[6] G. Carlsson (2009). Topology and data. *Bull. Amer. Math. Soc.*, **46**, 255–308.

[7] A. T. Casas (2019). Distributing persistent homology via spectral sequences. arXiv: 1907.05228.

[8] H. Edelsbrunner and J. Harer (2008). Persistent homology — A survey. In *Surveys on Discrete and Computational Geometry: Twenty Years Later*. Contemporary Mathematics, Vol. 453, (American Mathematical Society) pp. 257–282. https://www.maths.ed.ac.uk/~v1ranick/papers/edelhare.pdf.

[9] H. Edelsbrunnera and J. Harer (2010). *Computational Topology: An Introduction*, (American Mathematical Society). https://www.maths.ed.ac.uk/~v1ranick/papers/edelcomp.pdf.

[10] H. Edelsbrunner, D. Letscher and A. Zomorodian (2002). Topological persistence and simplification. *Discrete Comput. Geom.*, **28**, 511–533. https://link.springer.com/content/pdf/10.1007/s00454-002-2885-2.pdf.

[11] R. A. Fisher (1936). The use of multiple measurements in taxonomic problems. *Ann. Eugen.*, **7**(Part II), 179–188. https://archive.ics.uci.edu/ml/datasets/iris.

[12] R. Ghirst (2014). *Elementary Applied Topology* (Self Publication). https://www2.math.upenn.edu/~ghrist/notes.html.

[13] A. Goldfarb, J. B. Kwon, and T. Snider (2016). Detecting potential product segments using topological data analysis. *Working Paper*. https://www-2.rotman.utoronto.ca/~agoldfarb/TDA.pdf.

[14] B. Goldfarb (2018). The Mapper algorithm and its applications. *15th Annual Workshop on Topology and Dynamical Systems*. http://topology.nipissingu.ca/workshop2018/slides/Goldfarb-Data-Beamer.pdf.

[15] A. Hatcher (2002). *Algebraic Topology* (Cambridge University Press). https://pi.math.cornell.edu/~hatcher/AT/ATpage.html.

[16] L. Kaufman, and P. J. Rousseeuw (1990). Partitioning around medoids (program PAM). In *Finding Groups in Data: An Introduction to Cluster Analysis*. Wiley Series in Probability and Statistics, John Wiley & Sons, Inc., Hoboken, NJ, USA, pp. 68–125.

[17] A. Khaled, K. Minkyu and L. Jeongkyu (2017). Extracting knowledge from the geometric shape of social network data using topological data analysis. *Entropy*, **19**(7). https://www.mdpi.com/1099-4300/19/7/360/htm.

[18] J. M. Kleinberg (2002). An impossibility theorem for clustering. *Advances in Neural Information Processing Systems*, pp. 446–453. https://www.cs.cornell.edu/home/kleinber/nips15.pdf.

[19] P. Y. Lum *et al.* (2013). Extracting insights from the shape of complex data using topology. *Sci. Rep.*, **3**, 1236. https://www.nature.com/articles/srep01236.

[20] M. Mahajan, P. Nimbhorkar, and K. Varadarajan (2009). The planar k-means problem is NP-hard. In *International Workshop on Algorthims and Computation*, Lecture Notes in Computer Science, Vol. 5431, (Springer, Berlin, Heidelberg), pp. 274–285.

[21] S. Moro, P. Cortez, and P. Rita (2014). A data-driven approach to predict the success of bank telemarketing. *Decis. Support Sys.*, **62**, 22–31. https://archive.ics.uci.edu/ml/datasets/bank+marketing.

[22] M. Nicolau, A. J. Levine, and G. Carlsson (2011). Topology based data analysis identifies a subgroup of breast cancers with a unique mutational profile and excellent survival. *Proc. Nat. Acad. Sci. USA*, **108**(17), pp. 7265–7270. https://www.ncbi.nlm.nih.gov/pmc/articles/PMC3084136/.

[23] R. Ostrovsky, Y. Rabani, L. Schulman, and C. Swamy (2006). The effectiveness of Lloyd-type methods for the k-means problem, *Proceedings of the 47th Annual IEEE Symposium on Foundations of Computer Science (FOCS'06)*, (IEEE), pp. 165–174. https://web.cs.ucla.edu/~rafail/PUBLIC/76.pdf.

[24] F. Pedregosa, G. Varoquaux, A. Gramfort, V. Michel, and B. Thirion (2011). Scikit-learn: Machine learning in Python. *J. Mach. Learn. Res.*, **12**(85), 2825–2830. https://jmlr.csail.mit.edu/papers/v12/pedregosa11a.html.

[25] M. Piekenbrock and J. A. Perea (2021). Move Schedules: Fast persistence computations in sparse dynamic settings. *arXiv*: https://arxiv.org/abs/2104.12285.

[26] N. Saul and C. Tralie (2019). Scikit-TDA: Topological data analysis for Python. https://github.com/scikit-tda/scikit-tda.

[27] D. Sculley (2010). Web scale K-means clustering. *Proceedings of the 19th International Conference on World Wide Web*. https://www.eecs.tufts.edu/~dsculley/papers/fastkmeans.pdf.

[28] G. Singh, F. Mémoli and G. Carlsson (2007). Topological methods for the analysis of high dimensional data sets and 3D object recognition. *Eurographics Symposium on Point-Based Graphics*. https://research.math.osu.edu/tgda/mapperPBG.pdf.

[29] C. Sotiriou, S.-Y. Neo, L. M. McShane *et al.* (2003). Breast cancer classification and prognosis based on gene expression profiles from a population-based study. *Proc. Natl. Acad. Sci. USA*, **100**(18), 10393–10398. https://www.pnas.org/content/pnas/100/18/10393.full.pdf.

[30] L. Yuanhong, D. Ming, and H. Jing (2007). A Gaussian mixture model to detect clusters embedded in feature subspace. *Commun. Inf. Syst.*, **7**(4), 337–352. https://projecteuclid.org/journals/communications-in-information-and-systems/volume-7/issue-4/A-Gaussian-Mixture-Model-to-Detect-Clusters-Embedded-in-Feature/cis/1211574970.full.

[31] H. Wagner, P. Dlotko, and M. Mrozek (2012). Computational topology in text mining. In *Computational Topology in Image Context*, Lecture Notes in Computer Science. M. Ferri, P. Frosini, C. Landi, A. Cerri, B. Di Fabio (eds.), Vol. 7309, (Springer, Berlin, Heidelberg), pp. 68–78. https://www2.math.upenn.edu/~dlotko/textMining.pdf.

[32] Wikipedia, *Cluster Analysis*. Wikipedia. https://en.wikipedia.org/wiki/Cluster_analysis.

[33] Wikipedia, *Single-linkage Clustering*. Wikipedia. https://en.wikipedia.org/wiki/Single-linkage_clustering.

Chapter 9

Qualitative Queries with Fuzzy Techniques

Konstantinos A. Raftopoulos* and Nikolaos K. Papadakis†

Hellenic Military Academy, Vari, Attica, Greece

* *raftop@cs.ucla.edu*
† *npapadakis@sse.gr*

This chapter elaborates on techniques for extending traditional query schemes by introducing qualitative attributes in a semantic context. Fuzzy set theory and elements of *Mathematical Psychology* are combined in a way that new possibilities emerge in qualitative query posing and answering. A query is seen as a set-valued function from the attributes domain into the power set of all possible responses. For each attribute, a *relaxation* technique is introduced, based on the local *set-valued density* of the answer space. The proposed technique is based on a semantic treatment enabling an optimal query scope. By extending standard fuzzy set algebra operations, query attributes are combined into a single fuzzy measurement of accuracy by means of which the query scope is dynamically adjusted. New capabilities in query answering are demonstrated, allowing for quantitative responses to qualitative questions, not possible with traditional query schemes.

1. Introduction

Traditional query processing accepts precisely specified queries and
provides exact answers, thus requiring users to fully understand the
problem domain and the database schema, returning limited or even
null information if an exact answer is not available. Typically, a user
is not aware of the database schema and/or has only a vague impres-
sion of the desired answer, thus the traditional query handling proves
to be inefficient in this case. In imprecisely stated queries, the term
query relaxation is used to describe the process of adjusting the query
scope by enlarging the search range or the answer scope to include
additional information. Enlarging and/or shrinking a query scope
can be accomplished either by enlarging the scope of the query itself
or the query's answer set, as illustrated in Figure 1. Without loss of
generality, one can initially assume a single attribute and see it as a
mapping from the attribute's domain into the power set of possible
retrievals. A query on a single attribute instance a_0 will map this
instance to the subset A_0 of all tuples that take the value a_0 at the
attribute a. This set A_0 is the *exact answer* to the query on instance
a_0. One can then say that $q(a_0) = A_0$ or A_0 is the image of instance
a_0 (of attribute a) under query q.

Seeing a query as a mapping on D (the domain in which the
attribute a takes values) allows a metric in D to be defined; a semantic metric the properties of which and the difficulties in its definition
will be examined hereinafter. Given this *semantic metric* defined on

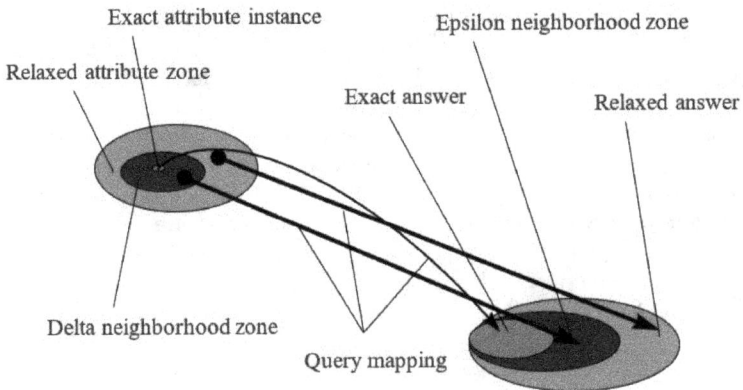

Figure 1.   Query as a set function and the respective relaxation topology.

the Cartesian product $D \times D$ into the set $\mathbb{R}$ (of reals), a distance is induced between two attribute instances in D and equivalently a neighborhood around a given instance a_0 is also defined. Thus, an r-neighborhood with center a_0 and radius r denoted by $\eta(a_0, r)$ is defined as the set of all the instances in D that have distance less than r from a_0. Now, instead of querying on the exact attribute instance a_0, one can relax the query in an $\eta(\alpha_0, r)$ neighborhood *around* a_0 and thus take a wider answer set which will include the exact answer set as a subset. This new answer set will be denoted by A_r, and therefore A_0, is a subset of A_r.

A_r is the image of $\eta(a_0, r)$ under query q, the relaxed answer, the wider answer one gets after relaxing the attribute value requirements. A_r is also the union of answer sets of all instances within range r from a_0, including A_0.

Let us now assume a measure defined in the power set of all possible answer sets R, so every set of tuples in R is measurable to a real value. One can then compare two answer sets A_0 and A_ϵ and measure their difference. Suppose $\eta(a_0, \delta)$ a δ neighborhood around a_0 mapped through query q to the answer set A_ϵ and $\eta(a_0, \delta'), \delta' > \delta$ a wider neighborhood with the same origin mapped to A'_ϵ through the same query. Knowledge on the difference of $\delta' - \delta$ should be enough to bound the difference $|A_\epsilon'| - |A_\epsilon|$ which is the difference of the measures of the two different answer sets after and before the relaxation step on the same query. It also permits adjusting the semantic relaxation to the desired value. It would have been desirable of course to perform the reverse procedure as well, that is, to apply the necessary relaxation to produce a specific pattern on the answer set or even to produce an enlargement of the answer set due to the insertion of additional associative information. This would produce a new query given the relaxed answer set, thus helping users in forming queries that produce answers they want. The circumstances under which this can take place will be examined hereinafter in the context of a nearness measure and cooperative query answering.

The rest of this chapter is organized as follows. A brief introduction to the general theory of measure and fuzzy sets is immediately provided and key elements from cognitive science and *Mathematical Psychology* regarding measures in semantic spaces are also covered. A proposed scheme for dynamic query scope adjustment is then introduced by means of a new semantic distance in both totally ordered and partially order domains. Finally, as an application scenario, an extension to standard fuzzy algebra operations is proposed, demonstrating significant improvements over traditional querying schemes. Parts of this chapter have appeared elsewhere, however this is the first attempt to combine fuzzy sets concepts with techniques from *Mathematical Psychology* and *Theory of Measurement and Scaling* toward a theoretical foundation of automatic query scope adjustment.

2. Measure

A short introduction to the theory of measure is attempted in this section. The reader is referred to Refs. [1,2] for further details on this subject. *Measure* provides an abstraction from the usual metric function that is often used in real life for measuring distances. It can be used as a quantitative tool that captures all the important aspects of the usual metric, transferring them to any arbitrary space with a minimum σ-algebra structure. It can be defined in arbitrary spaces carrying a minimal structure, as will be formally defined. It is a set function that maps sets to real numbers. All sets in the structure can be measured and become measurable sets. In order for a measure to be defined, apart from the implied structure of the space itself, the additiveness property is necessary as well, which implies the monotonicity of the measure function under addition.

Definitions will now be provided together with important properties.

Definition 1. A family $\mathcal{A}$ of subsets of a set X is called *algebra* in X iff it satisfies the following conditions:

$$X \in \mathcal{A}, \tag{1}$$

$$\text{if } A \in \mathcal{A}, \quad \text{then } A^c \equiv X \backslash A \in \mathcal{A}, \tag{2}$$

$$\text{if } A_1, A_2, \ldots, A_n \in \mathcal{A}, n \in N, \quad \text{then } \bigcap_{i=1}^{n} A_i \in \mathcal{A}, \tag{3}$$

meaning $\mathcal{A}$ is closed under complement and finite intersections.

Definition 2. A family $\mathcal{A}$ of subsets of an arbitrary set X is called *σ-algebra* in X iff it satisfies the following conditions:

$$X \in \mathcal{A}, \tag{4}$$

$$\text{if } A \in \mathcal{A}, \quad \text{then } X \backslash A \in \mathcal{A}, \tag{5}$$

$$\text{if } A_n \in \mathcal{A} \; \forall \, n \in N, \quad \text{then } \bigcap_{n=1}^{\infty} A_n \in \mathcal{A}, \tag{6}$$

meaning $\mathcal{A}$ is closed under complement and countable intersections. Let X be a set and $\mathcal{A}$ an σ-algebra in X.

Definition 3. A set function $\mu : \mathcal{A} \to [0, +\infty]$ is called a countably additive or σ-additive measure iff it satisfies the following properties:

$$\mu(\emptyset) = 0. \tag{7}$$

If A_n is a series of disjoint sets in $\mathcal{A}$, then

$$\mu \left(\bigcup_{n=1}^{\infty} A_n \right) = \sum_{n=1}^{\infty} \mu(A_n). \tag{8}$$

The pair $(X, \mathcal{A})$ is called measurable space and μ is the measure. Sets in $\mathcal{A}$ are called $\mathcal{A}$-measurable or simply measurable sets. Let $(X, \mathcal{A}, \mu)$ be a measurable space. If $A, B \in \mathcal{A}$ and $A \subset B$, then $\mu(A) \leq \mu(B)$ (μ is a single-valued set function). If in addition $\mu(A) < +\infty$, then $\mu(B \backslash A) = \mu(B) - \mu(A)$. Let $(X, \mathcal{A}, \mu)$ be a measurable space. μ is called

1. finite measure if $\mu(X) < +\infty$,
2. probability measure if $\mu(X) = 1$,
3. σ-Finite if there is a series A_n in $\mathcal{A}$ such that $\cup_{n=1}^{\infty} A_n = X$ and $\mu(A_n) < +\infty, \forall \, n$.

3. Fuzzy Sets

A short introduction to the theory of fuzzy sets is attempted in this section. The reader is referred to Refs. [3–8] for further information. Fuzzy set theory may be viewed as an attempt at developing body of concepts and techniques for dealing in a systematic way with a type of imprecision which arises when the boundaries of a class of objects are not sharply defined. Among the very common examples of such classes are the classes of big cars, narrow streets, short sentences, and funny jokes. Membership in such classes or, as they are suggestively called, fuzzy sets, is a matter of degree rather than an all or nothing proposition. Thus, informally, a fuzzy set may be regarded as a class in which there is a graduality of progression from membership to non-membership or more precisely in which an object may have a grade of membership intermediate between unity (full membership) and zero (nonmembership). In this perspective, a set in the conventional mathematical sense of the term may be viewed as a degenerate case of a fuzzy set, that is, a fuzzy set which admits of only two grades of membership: unity and zero.

Clearly, most classes of objects one encounters in real world are fuzzy sets in the informal sense defined above. Yet, the major focus of attention in mathematics, logic, and the hard sciences has been and continue to be centered on classes which are sets in the traditional sense. In the main, this is due to the misconception that fuzziness is a form of randomness and as such can be adequately treated by the tools provided by probability theory. However, as one develops a better understanding of the different varieties of imprecision, the following aspects are becoming increasingly clear:

- Fuzziness is fundamentally different from randomness.
- Fuzziness plays a much more basic role in human cognition than randomness.

To deal with fuzziness effectively, one may have to abandon many long-held beliefs and attitudes, and develop radically new conceptual frameworks for the analysis of humanistic as well as mechanistic systems. In the context of this chapter, which is effective and clever query answering, fuzzy set theory provides a useful tool for quantifying uncertainty in a way that is compatible to human cognition.

3.1. Measuring beliefs with membership functions: Possibility versus probability

The concept of fuzzy sets will be used now to measure semantic affinity in a way consistent to the general measure theory provided above. The key element in defining a fuzzy set is the membership function, a probabilistic function defined on the whole universe of discourse and taking values in $[0, 1]$. It is safe to think of it as a property the space acquires through this function. The membership function maps each element in the space of discourse to a real value in $[0, 1]$; this value quantifies one's grade of belief that this property is possessed by the particular element. This fundamental role the membership function plays in the concept of a fuzzy set often gives rise to confusion between fuzzy sets and probability. Fuzzy sets and probability are two conceptually orthogonal approaches toward describing uncertainty.

Probability is concerned with occurrence of well-defined events; fuzzy sets deal with classes of objects with unclearly defined boundaries. Let us quote: "Randomness has to do with uncertainty concerning membership or non membership of an object in a non fuzzy set, while fuzziness has to do with classes in which there may be grades of membership intermediate between full membership and non membership." Bezdek's discussion makes this point very clear. The origin of probability is distinct from that of fuzzy sets. We can see that by examining their values before and after a relevant experiment.

Suppose that before an experiment, the *a priori* probability of an outcome (x) to occur is equal to $P(x)$ and the respective membership value is $A(x)$. Obviously, the meaning of these two numbers is very different. After the experiment involving x, the probability turns out to be either 1 or 0 because x becomes the result of the experiment or not. The same experiment does not affect the membership value; after the experiment, it is still equal to $A(x)$. The way of perceiving the concept is left unchanged. The orthogonallity of the notions of probability and possibility does not exclude emergence of many quite different mixed situations; in practice, they can appear quite frequently.

Consider the statements:

P1: Temperature is high (fuzziness).
P2: Probability (temperature is 40) is 0.7 (probability).
P3: Probability (temperature is high) is low (?).

This last proposition involves both the uncertainties. While the notions of fuzziness and probability are orthogonal, this does not preclude them from any interaction. This phenomenon of symbiosis has been appreciated in early chapters that dealt with fuzzy probabilities, fuzzy expected values and fuzzy random variables [7]. Merging techniques and formalization models do exist, but are beyond the scope of this chapter.

3.2. *Fuzzy set algebra*

A fuzzy subset A of the universe of discourse U is characterized by a membership function $\mu_A : U \to [0, 1]$ or more generally into a lattice. $\mu_A(u)$ represents the grade of membership of u in A and has the following properties:

Containment:

$$A \subset B \Leftrightarrow \mu_A(u) \leq \mu_B(u), \quad \forall \, u \in U, \tag{9}$$

Complement:

$$A© \triangleq \int_U \frac{1 - \mu_A(u)}{u}, \tag{10}$$

Union:

$$A \cup B \triangleq \int_U \frac{\mu_A(u) \vee \mu_B(u)}{u}, \quad (\vee \text{ stands for max}), \tag{11}$$

Intersection:

$$A \cap B \triangleq \int_U \frac{\mu_A(u) \wedge \mu_B(u)}{u}, \quad (\wedge \text{ stands for min}), \tag{12}$$

Product:

$$AB \triangleq \int_U \frac{\mu_A(u)\mu_B(u)}{u}, \tag{13}$$

Power:

$$A^\alpha \triangleq \int_U \frac{(\mu_A(u))^\alpha}{u}. \tag{14}$$

Fuzzy set A in the universe U is the collection of all the elements in U mapped into the interval $[0, 1]$ of the real numbers through the membership function. The fuzzy set algebra is defined through the membership functions and is summarized as follows from above:

- A fuzzy set A contains B iff B maps each element of U to smaller real values in $[0,1]$ than A does.
- A union B is a new fuzzy set C with membership function defined as the pointwise maximum of the other two membership functions.
- A intersection B is a new fuzzy set C with membership function defined as the pointwise minimum of the other two membership functions.
- A product B is a new fuzzy set C with membership function defined as the pointwise product of the other two membership functions.
- Power is defined directly from product.

After introducing the concept of fuzzy sets, new logical operations are also required since the standard TRUE, FALSE duality of Boolean logic has been replaced with degrees of *Truthness* by means of membership functions. The traditional AND, OR and NOT logic operations are thus extended to that of *conjunction, disjunction* and *negation*, respectively, as follows.

Definition 4. If μ_{Ψ_1} and μ_{Ψ_2} are the membership functions of fuzzy sets Ψ_1 and Ψ_2, respectively, then the disjunction of two elements ψ_1 and ψ_2 in Ψ_1 and Ψ_2, respectively, is denoted with $\psi_1 \vee \psi_2$ and has a membership value equal to $\mu_{\psi_1 \vee \psi_2} = \max\{\mu_{\Psi_1}(\psi_1), \mu_{\Psi_2}(\psi_2)\}$.

Disjunction, therefore, is the extension of logical operation OR since in the case of membership pairs $(0,1), (1,0), (1,1)$ and $(0,0)$, their disjunction gives memberships $1, 1, 1$ and 0, respectively. Similarly, the extension of logical AND is called *conjunction* defined as follows.

Definition 5. If μ_{Ψ_1} and μ_{Ψ_2} are the membership functions of fuzzy sets Ψ_1 and Ψ_2, respectively, then the conjunction of two elements ψ_1 and ψ_2 in Ψ_1 and Ψ_2, respectively, is denoted with $\psi_1 \wedge \psi_2$ and has a membership value equal to $\mu_{\psi_1 \wedge \psi_2} = \min\{\mu_{\Psi_1}(\psi_1), \mu_{\Psi_2}(\psi_2)\}$.

In the case of membership pairs $(0,1), (1,0), (1,1)$ and $(0,0)$, their *conjunction* gives memberships $0, 0, 1$ and 0, respectively, indeed equivalent to the logical AND operation in this case. Lastly, the traditional NOT operation is extended to that of *negation* as follows.

Definition 6. If μ_Ψ is a membership function of fuzzy set Ψ, then the *negation* of $\psi \in \Psi$ is denoted with $\rceil\psi$ and has membership value equal to $\mu_{\rceil\psi} = 1 - \mu_\Psi(\psi)$.

4. Measuring Semantic Spaces with Membership Functions

The field of fuzzy set theory has, since its inception by Zadeh, aroused controversy mainly due to opposition by those observers who believe that probability theory is the only means of modeling uncertainty. Yet, even among fuzzy set theorists themselves, controversy exists, especially regarding the measurement of fuzziness and the properties of the resulting membership functions. Here, the results of an empirical study about modeling semantic measurements through fuzzy set membership functions are discussed [9].

First, let us define the terms *subjective* and *objective* of an attribute. The former refers to an attribute A which is unambiguously either possessed or not possessed by any object j in a domain of discourse J. The latter refers to any attribute whose linguistic definition contains ambiguities. From the very fact that individuals may subjectively decide the amount A possessed by an object j in J, we see that the value of the membership function of j in the set induced in J by A (the fuzzy set representation of the attribute A), which we will for simplicity call set A, is not objective in nature. One therefore faces the hardship of assigning numbers to subjective perceptions. The construction of such a representation is the province of *mathematical psychology* utilizing techniques from the *theory of measurement and scaling* [1,2,10]. Such techniques will also be used hereinafter for measuring subjective attributes for individual subjects. These can be easily extended for measuring group membership functions as well.

4.1. Theory of measurement and scaling

Definition 7. Given a domain J, define the weak order $\prec$ in J so that $\theta_1 \prec \theta_2, \forall\, \theta_1, \theta_2 \in \Theta$ if an observer judges that j_1 is at least as A as j_2 is or it is at least as true that j_1 is A as it is that j_2 is A or j_1 is at least as large as j_2 with respect to being A. The system $\langle \Theta, \prec \rangle$ defined above will be called a multivalued membership structure.

This definition does not imply bounds on the membership function of a subjective set, so they must be explicitly added.

Definition 8. A membership structure $\langle \Theta, \prec \rangle$ is called "bounded" if there exist elements j_m and j_M such that $j_M \prec j$ and $j \prec j_m$ for all j in J.

Here, j_M is an object which is judged by the subject whose membership function is being constructed as "definitely being A" and j_m is an object which the subject judges as definitely not being A (these judgments interpreted as meaning that j_M has maximum membership in the set representation of A and j_m has minimum membership).

It is also required that the subject is able in general to compare any pair of intervals specified by a total of four points in J in order to establish a *weak order* of intervals in J. The weak order $\prec'$ permits the statement that j_2 is A-er than j_1 by at least as much as j_4 is A-er than j_3:

$$\theta_2 \theta_1 \prec' \theta_4 \theta_3, \quad \forall\, \theta_1, \theta_2, \theta_3, \theta_4 \in \Theta. \tag{15}$$

Definition 9. A bounded multivalued membership structure $\langle \Theta, \prec \rangle$ for which the intervals $j_i j_j$ in J can be weakly ordered by an ordering $\prec'$ for all j_i, j_j in J will be termed difference comparable and denoted $\langle \Theta \times \Theta, \prec' \rangle$.

Unlike the comparison defined above, this last definition is not a topological one about $\prec$ in J but an algebraic one and induces an *algebraic-difference structure*.

Now, let us assume that J is *order dense* with respect to $\prec$ which means that it can be represented by a continuum associated numerical domain $X(J)$. The introduction of $X(J)$ serves the dual purpose of rendering the graph of membership over X continuous and in accordance with the standard practice of fuzzy set literature, making membership a function of a numerical base variable rather than that of an empirical object. When a subject's perception of the attribute A as applied to an order-dense domain J satisfy the axioms of an algebraic-difference structure, then the following *numerical represen-tation* of membership is implied [2,5].

Theorem 1 (Representation theorem). *Let J be a set which is order dense, with $\langle \Theta, \prec \rangle$ a difference comparable bounded multivalued membership structure such that $\langle \Theta \times \Theta, \prec \rangle$ is an algebraic-difference structure. Then there exists a bounded real-valued function denoted m on J such that for all j_1, j_2, j_3, j_4 in J :*

$$\theta_2 \prec \theta_1 \qquad \Leftrightarrow \qquad \mu(\theta_2) \le \mu(\theta_1), \qquad (16)$$

$$\theta_2 \theta_1 \prec' \theta_4 \theta_3 \qquad \Leftrightarrow \qquad \mu(\theta_2) - \mu(\theta_1) \le \mu(\theta_4) - \mu(\theta_3). \quad (17)$$

The scale resulting from this theorem will be interpreted as the membership of j in A. $\langle \Theta \times \Theta, \prec' \rangle$ is now represented with a homomorphic numerical structure $\langle \mu, \le \rangle$ which can be shown as unique up to a positive linear transform.

Theorem 2 (Uniqueness theorem). *If m' is another function satisfying the above, then $m'(j) = c_1 m(j) + c_2, c_1 > 0$, i.e. m is on an interval scale.*

Recall that in accordance to *interval scale*, differences between values can be quantified in absolute but not relative terms and any zero is merely arbitrary, e.g. dates are measured on an interval scale since differences can be measured in years, but a ratio of dates has no meaning. It is compared to ordinal and ratio scales. The ratio scale has a fixed zero value and permits comparison of differences, e.g. although time itself cannot be measured on a ratio scale, differences in time can, since it makes sense to talk of one pair of events being twice as far apart as another. In the ordinal scale, data are shown simply in accordance to some order in the absence of appropriate units of measurement, e.g. a squash ladder is an ordinal scale since one can say only that one competitor is better than another, but no by how much. Ratio is the strongest scale and ordinal the weakest.

Now, let us return to m. The numerical representation incorporates two parameters: an arbitrary origin and an arbitrary unit. So, any element in J may be assigned with membership value 0 and any other may be assigned with value 1. Each $\theta_- \in [\theta_m] \equiv I_0$ can therefore take on an arbitrary value B^L of the lower bound on μ and each $\theta_+ \in [\theta_M] \equiv I_1$ can take on B^U as an arbitrary value of

Membership function

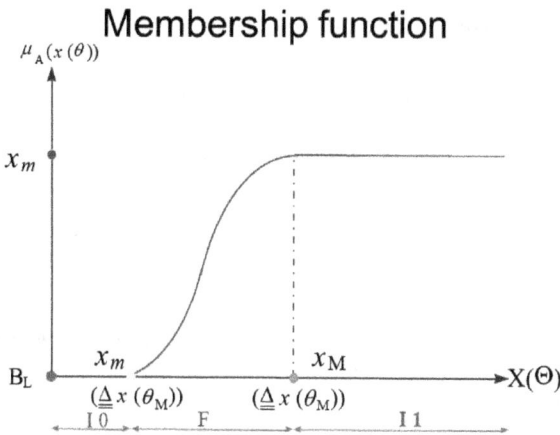

Figure 2. A membership function.

the upper bound on m. However, once two values B^L and B^U are chosen, then the membership values of all other elements in j are totally specified relative to this bounds. For example, the subjective attribute *Tall man* is denoted by $A = $ tall, $\theta = \{$men of various heights$\}$ and $X(\Theta) = [0, \infty]$ inches/centimeters. Function $m(x(\theta))$ is shown in Figure 2 as an increasing function that partitions $X(\Theta)$ into three regions I_0, F and I_1. This partitioning contrasts with that induced by an objective attribute, such as a man at least 60 inches tall, which splits $X(\Theta)$ into only the two subsets, $I_0 = [0, 60)$ and $I_1 = [60, \infty)$.

4.2. The question of extensive measurement

To this point, through representation and uniqueness theorems, it is shown that for J order dense, $m(j)$ is on at least interval scale. The question then arises whether membership can be on a stronger scale such as ratio. Ratio scales are common in the physical sciences but are rare in social science, where ordinal and interval scales are the rule. Since these sciences deal more directly with the human cognition, limitations in the use of ratio scales for semantics is implied as well. Extensive measurement is a common tool in physical sciences for obtaining ratio scales, e.g. for length and mass. However, for measuring psychological attributes, it has been applied successfully in only two cases, that of subjective probability and risk.

5.   Dynamic Query Scope Adjustment

Extending a query scope to a wider one when it contains insufficient information (relaxation) and shrinking a query scope to a finer one when it contains much optional information involve the mechanism of the dynamic query scope adjustment. A nonlinear measure of semantic distance will be introduced for guiding the query scope automatically via nearness measurements. Under this notion of measure, semantic distances are not measured evenly by a fixed scale. Assuming a query relaxation process is allowed only within a given semantic distance, measured by a larger scale, the given semantic distance corresponds to a smaller real-valued distance than measured by a smaller scale. Therefore, enlarging the scale leads the querying process to shrink the searching scope to find answers with higher accuracy; inversely by shrinking the scale, the process extends the search scope to accept answers with less accuracy.

Query relaxation can thus be controlled by a nearness measure, e.g. since many flights depart in the morning, using an enlarged scale can let 1 hour difference in departure time to be considered as a big gap and guide the system to search for alternative flights departing within 30 minutes; at the same time, the same 1 hour difference in departure time is considered as a small gap in late evening flights if a shrunken scale is used, thus guiding the system to search alternative flights departing within 2 hours. Traditionally, when measuring similarity between two concepts, their semantic distance is viewed symmetrically. However, from a *Cognitive science* point of view, a specific concept is closer to its generalization than inversely. In terms of the proposed similarity measure, an attribute value corresponding to fewer occurrences is semantically closer to one with more occurrences than inversely. The rational behind this assumption is that the attribute value instance with fewer occurrences depicts a specific notion with respect to another attribute instance with more occurrences. In general, to better support practical applications, the nearness measure in a given attribute domain should be based not only on the domain values but also on the object distribution on those values. Such a requirement justifies the use of nonlinear and asymmetrical measures.

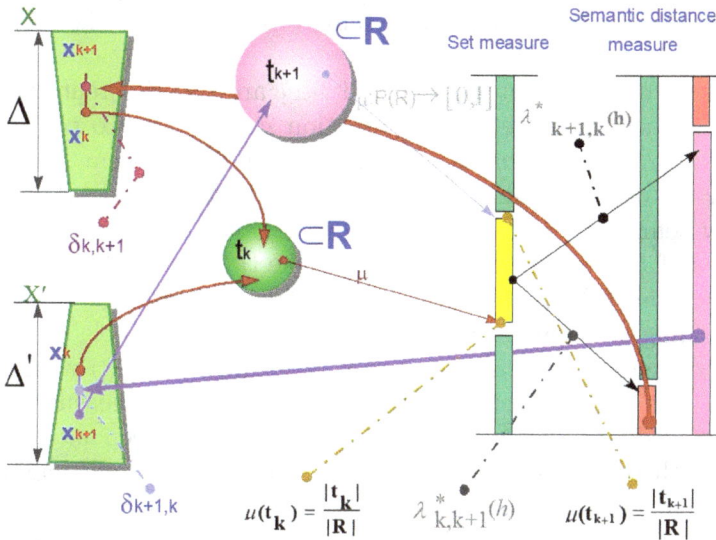

Figure 3. Semantic distance architecture.

5.1. *Semantic distance construction in a totally ordered domain*

The construction of a nearness measure that abides to the above principals is now described. A totally ordered domain X is assumed, among the members of which this measure will be defined. "Totally ordered" means it is possible to list its elements in ascending or descending order. Since the measure defined is not symmetric, for illustrative purposes, two different representations will be used for the same domain, one showing it in ascending order and the other in descending order as is also illustrated in Figure 3 to which the reader is referred during this construction.

The proposed measure have a stepwise definition. Let us choose two consecutive points in X, let them be x_k, x_{k+1} and denote their semantic distance as $\delta_{k,k+1}$ and $\delta_{k+1,k}$ (considering both directions). A query on a single attribute instance in X is seen as a mapping from this instance to its answer set. This means that a particular query q on x_k maps x_k to the set of objects that have value x_k at attribute x. This answer set is denoted with t_k (for simplicity, since one knows that attribute x is referenced). Thus, x_k is mapped to t_k

by means of query q and its consecutive point x_{k+1} in the ascending space X is mapped to t_{k+1}.

The two sets denoted as t_k and t_{k+1} are two different sets, two different values the set-valued function q takes in two consecutive points in X. Recall that q is viewed as a function from X into the power set of all possible responses R denoted by $P(R)$. A measure is now defined in $P(R)$ as the usual cardinality function mapping all subsets of R to their cardinality. To make this a probability measure (restrict the values in $[0,1]$), a linear transformation is introduced by dividing each value with the cardinality of all possible answers R. The new measure is called μ and it is defined on $P(R)$ with values into $[0,1]$. With μ well defined in $P(R)$, one can measure t_k and t_{k+1}, and define the semantic distance $\delta_{k,k+1}$ between x_k and x_{k+1} in X with respect to their relative measurement. In particular, one can calculate a weighted sum of the two measurements $\mu(t_k), \mu(t_{k+1})$ denoted as $\lambda^*_{k,k+1}$. This weighted sum is defined as

$$\lambda^*_{k,k+1} = \frac{h}{2}\mu(t_k) + \left(1 - \frac{h}{2}\right)\mu(t_{k+1}), \quad 0 \le h \le 1, \qquad (18)$$

and depending on the parameter h, it can be any point in the interval:

$$\left[\min\{\mu(t_k), \mu(t_{k+1})\}, \frac{\max\{\mu(t_k),\mu(t_{k+1})\} + \min\{\mu(t_k),\mu(t_{k+1})\}}{2}\right].$$

Note that $\lambda^*_{k+1,k}$ is not equal to $\lambda^*_{k,k+1}$ except in the case where $h = 1$. Now, already the distance from x_k to x_{k+1} can be defined as

$$\delta_{k,k+1} = \gamma\left(1 - \beta\left(\frac{1}{n} - \lambda^*_{k,k+1}\right)\right), \quad 0 \le \beta \le 1. \qquad (19)$$

In Figure 3, one can see λ^* mapping the length $I = \max\{m(t_k), m(t_{k+1})\} - \min\{m(t_k), m(t_{k+1})\}$ to the length I' that corresponds to its subinterval defined by the intermediate point $\lambda^*_{k,k+1}$. The difference between I' and $1/n$, where n is the cardinality of X (the domain), is magnified by β which is the parameter adjusting the impact of the attribute occurrence to their distance. This value, adjusted by base scale gamma (γ), provides a semantic distance between x_k and x_{k+1}. Base scale gamma (γ) can be calculated from the domain's diameter, where diameter is the distance between x_1 and x_n in X and denoted by Δ and Δ' for the ascending and descending arrangements, respectively. It is usually assumed to be unity unless there is a good reason against it. Base gamma can be calculated from

$$\sum_{k=0}^{n-2} \delta_{k,k+1} = \gamma \sum_{k=0}^{n-2} \left(1 - \beta\left(\frac{1}{n} - \frac{h}{2}\mu(t_k) - \left(1 - \frac{h}{2}\right)\mu(t_{k+1})\right)\right) = \Delta,$$

$$\sum_{k=0}^{n-2} \delta_{k+1,k} = \gamma \sum_{k=0}^{n-2} \left(1 - \beta\left(\frac{1}{n} - \frac{h}{2}\mu(t_{k+1}) - \left(1 - \frac{h}{2}\right)\mu(t_k)\right)\right) = \Delta'.$$

$$(20)$$

Recall that $\lambda^*_{k,k+1}$ is not equal to $\lambda^*_{k+1,k}$ which in terms of the distance definition in (19) means that $\delta_{k,k+1}$ is not equal to $\delta_{k+1,k}$, thus two different but symmetrical values in $[0,1]$ are assigned as distances from x_k to x_{k+1} and x_{k+1} to x_k, respectively, as is also shown in Figure 3. In other words, the piecewise distance just constructed is not symmetric, however this asymmetricity is justified as a cognitive science requirement explained later. To summarize, a stepwise nearness function was just constructed in the totally ordered domain set X with diameter Δ. Any pair of points x_i, x_j in X have now distance $\delta_{i,j}, \delta_{j,i}$ that can be calculated through

$$\delta_{i,j} = \sum_{k=1}^{j-1} \delta_{k,k+1}, \quad i < j,$$

$$\delta_{i,j} = 0, \quad \text{if } i = j, \qquad (21)$$

$$\delta_{j,i} = \sum_{k=1}^{j-1} \delta_{k+1,k}, \quad i < j.$$

It is trivial to see that δ induces a measure into X in the form of a nonlinear piecewise additive function.

5.2. *In partially ordered domains*

The above construction can be naturally extended to partially ordered domains. The difference is that one cannot order the domain in ascending or descending order and define the measure piecewise. The piecewise definition is strong since it ensures the additiveness property that is necessary for any formal measure definition (see above). The lack of a total ordering does not allow many options other than building a distance table that will directly contain distances between all pairs of points in X. In the (i, j) entry of this table, the distance from x_i to x_j will be kept. A convention found in fuzzy set theory usually provides a likeness relation for measuring semantic distances in partially ordered domains. This likeness relation is a symmetric table holding a real value between 0 and 1 in each entry. This value indicates the degree of one's belief that the two attribute values corresponding to the row and column of that entry are alike. Given such a table that can be constructed under the cognitive procedures covered earlier, one can adjust its entries, so it will form an asymmetric unlikeness measure or a measure of distance that takes into account the impact of the occurrences of the attribute values under measurement in the domain in question. If one, therefore, denotes with $\sigma_{i,j}$ an arbitrary entry of the likeness table, it can be modified to

$$d_{i,j} = (1 - \sigma_{i,j})H, \quad \text{where } H = 1 - \beta\left(\frac{1}{n} - \lambda^*_{\alpha_i, \alpha_j}\right), \quad 0 \le \beta \le 1$$

and

$$\lambda^*_{\alpha_i, a_j} = \frac{h}{2}\mu(t_{\alpha_\iota}) + \left(1 - \frac{h}{2}\right)\mu(t_{\alpha_j}), \quad 0 \le h \le 1.$$

m is the measure in the set of all possible answers R as defined in the totally ordered case, β and h play exactly the same role as above and t_v is again as defined above: the set of responses having value v at the attribute in question. This modified $d_{i,j}$ entry describes the distance table introduced for query scope adjustment. Note that unlike $\sigma_{i,j}$, it is asymmetric, modeling this way a natural tendency to *further distinguish similar concepts*.

5.3. *The adjustment mechanism*

At the first relaxation step, a query is *relaxed* to include attribute values inside a certain distance from the initial value. This distance is calculated as above depending on the ratio of occurrences of these values compared to occurrences of the current one.

For a pair of attribute values that show the same occurrence, the distance between them is calculated to $1/n$ where n is the cardinality of the domain space. If the neighboring attribute value has more occurrences, it is considered a generalization over the current value and the distance is calculated less than $1/n$, so it is getting closer, if it has less occurrences, it is considered a specialization over the current one and it is getting farther away by being assigned a distance more toward 1.

If the query relaxation step is a fixed value step, it will take more relaxation steps to include a specialized attribute in the answering set that it will need to go the other way as it is shown in Figure 4. The piecewise definition, combined with the total ordering of the domain in question (in the case of totally ordered domains) and the existing likeness relation (in the case of partially ordered domains), indicate that the *nearness measure described is also value-dependent.*

6. Fuzzy Algebra Extension for Gradual Logic Connectiveness

Up to now, the following techniques have been discussed:

1. calculating semantic distances for a single attribute,
2. relaxing the search, based on these distances,
3. building membership functions for modeling a property as a fuzzy subset of the universe of discourse for a single attribute.

Queries with multiple attributes are combined with the connectives of conjunction, disjunction and negation (Definitions 4–6) in Section 3.2. An application scenario will now be presented, showing a combination of the above techniques in extending traditional query answering. The reader is referred to Refs. [11,12] for similar approaches.

Suppose a query to an airliner database for "low price morning" flights. In processing this query, there are two domains involved: the "price" taking instances in dollars/euro and the "time of departure" with instances in time. "Low" is a property applied to the price universe of discourse and can be modeled as a fuzzy set in the price domain. The respective membership function can be constructed according to the above *representation theory* in Section 4.

Starting from an instance that is definitely low (membership 1) and relaxing the query to successive neighborhood zones around the

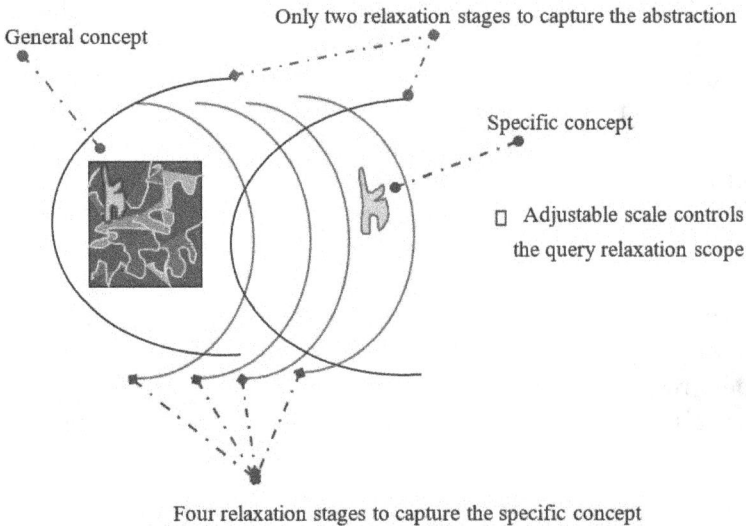

Figure 4.   The mechanism of adjusting the query scope in multiple asymmetric steps.

definite "low" values while measuring relaxation with values of the membership function, the price domain can be relaxed up to membership 0.7 and time domain up to membership 0.5. If one invokes the classic fuzzy set operation of conjunction to decide the membership of the marginal values in the new fuzzy set formed by the conjunction of the previous ones, the resulting membership value will equal 0.5, the minimum of previous memberships. This process results in "low price" relaxed to 0.3 from the definitely "low" and "morning flights" relaxed to 0.5 from the definitely "morning time". Had one relaxed the price to 0.8 or to any other value larger than 0.5, the conjunction fuzzy set would have the same membership function. It is then obvious that the way classic fuzzy set theory treats memberships, produced out of classic fuzzy set algebra, is not adequate for controlling the relaxation process. For this reason, a natural extension to the existing fuzzy set algebra, namely, the ∇-*connection* will be introduced to allow the construction of intermediate membership functions.

Definition 10. Let A and B be fuzzy sets with membership functions μ_A and μ_B, respectively. Then the ∇-connection of A and B is a new fuzzy set denoted by $A\nabla_\eta B$ having a membership function

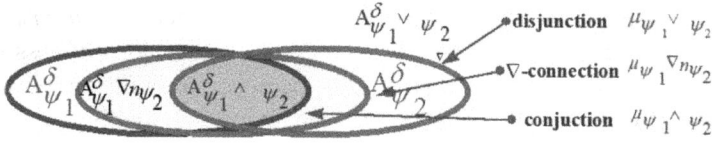

disjunction	$\max(\mu_{\psi_1} \cdot \mu_{\psi_2})$	$\mu_{\psi_1} \vee \psi_2$
∇-connection	$[\min(\mu_{\psi_1} \cdot \mu_{\psi_2}), \max(\mu_{\psi_1} \cdot \mu_{\psi_2})]$	$\mu_{\psi_1} \nabla_n \psi_2$
conjuction	$\min(\mu_{\psi_1} \cdot \mu_{\psi_2})$	$\mu_{\psi_1} \wedge \psi_2$

Figure 5. Fuzzy algebra extension to permit gradual logic connectiveness.

denoted by $\mu_{A\nabla_\eta B}$ and defined as

$$\mu_{A\nabla_\eta B} = \int_U \eta \max\{m_A(u), m_B(u)\} + \frac{(1-\eta)\min\{m_A(u), m_B(u)\}}{u},$$
(22)

where h is a parameter in $[0, 1]$ that adjusts the intermediate value the new membership function takes between min and max of μ_A and μ_B.

For different values of h, different membership functions will result from this operation. Let us now perform again the previous evaluation of relaxing the query about "low morning flights", first to a membership of 0.7 and then to a membership of 0.8 in price domain. In the first case, for $h = 0.3$, "low morning" query is relaxed to $0.3 * 0.7 + 0.7 * 0.5 = 0.56$, and in the second case, it is relaxed to $0.3 * 0.8 + 0.7 * 0.5 = 0.59$ which indeed provides the desired result since the relaxation in the second case is more strict specifying higher marginal values for the membership function. In Figure 5 the new operation of ∇-*connection* is shown graphically, capturing intermediate relaxation zones between the traditional min, max zones of the classic fuzzy set algebra.

7. Epilogue

A nearness measure in semantic spaces has been introduced, modeling a query as a set-valued function from the domain space into the power set of responses, giving rise to the concept of *query gradient*. Keeping the query *gradient* bounded, a relaxation process in the semantic space can be guided. The proposed relaxation technique combined with an extension to classic fuzzy set algebra allows for qualitative queries, thus improving traditional querying schemes.

References

[1] D. H. Krantz, R. D. Luce, P. Suppes, and A. Tversky (1971). *Foundations of Measurement*, Vol. 1 (Academic Press, New York).

[2] F. S. Roberts (1979). *Measurement Theory with Applications to Decision Making Utility and the Social Sciences* (Addison Wesley, Reading, MA).

[3] P. Bosc and H. Prade (1993). An introduction to fuzzy set and possibility theory-based approaches to the treatment of uncertainty and imprecision in data base management systems. *Proceedings of the Workshop on Uncertainty Management in Information Systems: From Needs to Solutions*, Vols. 2–5 (Avalon Santa Catalina, California).

[4] J. C. Bezdek (1993). Fuzzy models — What are they and why? *IEEE Trans. Fuzzy Syst.*, **1**, 1–6.

[5] J. C. Bezdek (1993). The thirsty traveler visits Gamont: A rejoinder to fuzzy models — What are they and why? *IEEE Trans. Fuzzy Syst.*, **2**, 43–45.

[6] J. C. Bezdek (1994). Special issue — Fuzziness vs probability — The n-th round. *IEEE Trans. Fuzzy Syst.*, **2**(1), 1–42.

[7] H. Kwakernaak (1979). Fuzzy random variables: I definitions, II algorithms and examples in the discrete case. *Inform. Sci.*, **15**, 253–278.

[8] L. A. Zadeh (1968). Probability measures of fuzzy events. *J. Math. Anal. Appl.*, **23**, 421–427.

[9] A. M. Norwich and I. B. Turksen (1993). A model for the measurement of membership and the consequences of its emperical implementation. In *Readings in Fuzzy Sets for Intelligent Systems* (Morgan Kaufman, San Francisco, CA, USA).

[10] L. Narens (2003). *Theories of Meaningfulness*. Scientific Psychology Series (Erlbaum Associates, Mahwah, NJ).

[11] W. W. Chu and Q. Chen (1992). Neighborhood and associative query answering. *J. Intell. Inf. Syst.*, **1**(3–4), 355–382.

[12] K. Hirota (1990). Concepts of probabilistic sets. *Fuzzy Sets Syst.*, **17**, 249–275.

Chapter 10

General DKH Contractions in Metric Spaces

Mihai Turinici

A. Myller Mathematical Seminar; Alexandru Ican Cuza University, Iaşi, Romania

mturi@uaic.ro

A fixed point theorem is established for a general class of DKH contractions over metric spaces. Technical connections with some statements in the area obtained by Du *et al.* are also given.

1. Introduction

Let X be a nonempty set. Call the subset Y of X, *almost singleton* (*asingleton*) provided [$y_1, y_2 \in Y$ implies $y_1 = y_2$], and *singleton* if, in addition, Y is nonempty; note that in this case, $Y = \{y\}$ for some $y \in X$. Take a metric $d : X \times X \to R_+ := [0, \infty[$ over X; the couple (X, d) will be referred to as a *metric space*. Further, let $T \in \mathcal{F}(X)$ be a self-map of X. [Here, for each couple A, B of nonempty sets, $\mathcal{F}(A, B)$ denotes the class of all functions from A to B; when $A = B$, we write $\mathcal{F}(A)$ in place of $\mathcal{F}(A, A)$]. Denote $\text{Fix}(T) = \{x \in X; x = Tx\}$; each point of this set is referred to as *fixed* under T. The determination of such points is to be performed in the following context, comparable with the one in Rus [24, Chapter 2, Section 2.2]:

(**pic-1**) We say that T is a *Picard operator* (modulo d) if, for each $x \in X$, the iterative sequence $(T^n x; n \geq 0)$ is d-Cauchy: for each $\varepsilon \in R_+^0 :=]0, \infty[$, there exists an index $k = k(\varepsilon)$ such that $k \leq n \leq m$ imply $d(x_n, x_m) < \varepsilon$.

(**pic-2**) We say that T is a *strong Picard operator* (modulo d) if, for each $x \in X$, $(T^n x; n \geq 0)$ is d-convergent with $\lim_n (T^n x) \in \mathrm{Fix}(T)$.

(**pic-3**) We say that T is *fix-asingleton* if $\mathrm{Fix}(T)$ is an asingleton; and *fix-singleton*, provided $\mathrm{Fix}(T)$ is a singleton.

Concerning the existence and uniqueness of such points, a basic result, referred to as *Banach fixed point theorem* (B-fpt) may be stated as follows. Call the self-map T, $(d; \alpha)$-*contractive* (where $\alpha \geq 0$), if

(con) $d(Tx, Ty) \leq \alpha d(x, y)$, for all $x, y \in X$.

Theorem 1. *Suppose that T is $(d; \alpha)$-contractive for some $\alpha \in [0, 1[$. In addition, let X be d-complete. Then, T is strong Picard (modulo d) and fix-singleton.*

This result, established in 1922 by Banach [3], found some important applications to the operator equations theory. Consequently, a multitude of extensions for (B-fpt) were proposed. From the perspective of this exposition, the set implicit ones are of interest. Denote, for $x, y \in X$,

$Q_1(x, y) = d(x, Tx)$, $Q_2(x, y) = d(x, y)$, $Q_3(x, y) = d(x, Ty)$,
$Q_4(x, y) = d(Tx, y)$, $Q_5(x, y) = d(Tx, Ty)$, $Q_6(x, y) = d(y, Ty)$,
$\mathcal{Q}(x, y) = (Q_1(x, y), Q_2(x, y), Q_3(x, y), Q_4(x, y), Q_5(x, y), Q_6(x, y))$.

Then, the underlying contractions may be written as

(i-s-con) $\mathcal{Q}(x, y) \in \Upsilon$ for all $x, y \in X$,

where $\Upsilon \subseteq R_+^6$ is a (nonempty) subset. In particular, when Υ is the zero section of a certain function $\Phi : R_+^6 \to R$, i.e.

$\Upsilon = \{(t_1, \ldots, t_6) \in R_+^6; \Phi(t_1, \ldots, t_6) \leq 0\}$,

the implicit contractive condition above has the functional form (i-f-con) $\Phi(\mathcal{Q}(x,y)) \leq 0$ for all $x, y \in X$.

Concerning the "genuine" implicit case, some recent contributions in the area may be found in Akkouchi [1], Berinde and Vetro [4], or Nashine *et al.* [19]. We stress that in almost all chapters based on such techniques — including the ones we just quoted — it is claimed that their starting point is represented by the 1999 contribution due to Popa [21]. Unfortunately, this claim is not true: fixed point results based on implicit techniques were obtained more than two decades ago in Turinici [27,28]. But, we must note that some partial aspects of the set-implicit theory (as described above) have been discussed in the (classical by now) 1969 Meir–Keeler fixed point principle [15]; subsequently refined by Matkowski [14]. On the other hand, for the explicit case, some basic results were obtained in Boyd and Wong [6], Matkowski [12], Leader [11], Piticari [20, Chapter II] and Reich [22]; see also the survey chapter by Rhoades [23].

Having these precise, it is our aim in the following to give some fixed point results — involving six-dimensional contractive conditions — for self-maps acting upon metric spaces. Their basic tool is a lot of strong Matkowski admissible criteria including the one in Altman [2]. Then, as a by-product of these, one gets a simplified proof of the related statement in Du *et al.* [8]. Further extensions of the obtained facts to relational structures will be discussed elsewhere.

2. Dependent Choice Principle

Throughout this exposition, the axiomatic system in use is Zermelo–Fraenkel (ZF), as described by Cohen [7, Chapter 2]. The notations and basic facts to be considered in this system are more or less standard. Some important ones are discussed below.

(A) Let X be a nonempty set. By a *relation* over X, we mean any (nonempty) part $\mathcal{R} \subseteq X \times X$; then, $(X, \mathcal{R})$ will be referred to as a *relational structure*. For simplicity, we sometimes write $(x, y) \in \mathcal{R}$ as $x\mathcal{R}y$. Note that $\mathcal{R}$ may be regarded as a mapping between X and $\exp[X]$ (= the class of all subsets in X). In fact, denote

$$X(x, \mathcal{R}) = \{y \in X; x\mathcal{R}y\} \text{ (the *section* of } \mathcal{R} \text{ through } x), \quad x \in X;$$

then, the desired mapping representation is $[\mathcal{R}(x) = X(x, \mathcal{R}); x \in X]$.

A basic example of relational structure is to be constructed as follows. Let $N = \{0, 1, \dots\}$ be the set of *natural* numbers endowed with the usual addition and (partial) order; note that

$(N, \le)$ is well ordered: any (nonempty) subset of N has a first element.

Further, denote, for $p, q \in N$, $p \le q$,

$$N[p, q] = \{n \in N; p \le n \le q\}, \quad N]p, q[= \{n \in N; p < n < q\},$$
$$N[p, q[= \{n \in N; p \le n < q\}, \quad N]p, q] = \{n \in N; p < n \le q\},$$

as well as, for $r \in N$,

$$N[r, \infty[= \{n \in N; r \le n\}, \quad N]r, \infty[= \{n \in N; r < n\}.$$

By definition, $N[0, r[= N(r, >)$ is referred to as the *initial interval* (in N) induced by r. Any set P with $P \sim N$ (in the sense: there exists a *bijection* from P to N) will be referred to as *effectively denumerable*. In addition, given some natural number $n \ge 1$, any set Q with $Q \sim N(n, >)$ will be said to be *n-finite*; when n is generic here, we say that Q is *finite*. Finally, the (nonempty) set Y is called (at most) *denumerable* iff it is either effectively denumerable or finite.

Let X be a nonempty set. By a *sequence* in X, we mean any mapping $x : N \to X$, where $N = \{0, 1, \dots\}$ is the set of *natural* numbers. For simplicity, it will be useful to denote it as $(x(n); n \ge 0)$ or $(x_n; n \ge 0)$; moreover, when no confusion can arise, we further simplify this notation as $(x(n))$ or (x_n), respectively. Also, any sequence $(y_n := x_{i(n)}; n \ge 0)$ with

$(i(n); n \ge 0)$ is strictly ascending (hence, $i(n) \to \infty$ as $n \to \infty$)

will be referred to as a *subsequence* of $(x_n; n \ge 0)$. Note that, under such a convention, the relation "subsequence of" is *transitive*, i.e.,

(z_n) = subsequence of (y_n) and (y_n) = subsequence of (x_n) imply (z_n) = subsequence of (x_n).

(B) Remember that an outstanding part of (ZF) is the *Axiom of Choice* (AC); which, in a convenient manner, may be written as

(AC) For each couple (J, X) of nonempty sets and each function, $F : J \to \exp(X)$, there exists a (selective) function, $f : J \to X$, with $f(\nu) \in F(\nu)$, for each $\nu \in J$.

(Here, $\exp(X)$ stands for the class of all nonempty elements in $\exp[X]$). Sometimes, when the index set J is denumerable, the existence of such a selective function may be determined by using a weaker form of (AC) called *Dependent Choice* principle (DC). Call the relation $\mathcal{R}$ over X *proper* when

$$(X(x, \mathcal{R}) =)\mathcal{R}(x) \text{ is nonempty for each } x \in X.$$

Then, $\mathcal{R}$ is to be viewed as a mapping between X and $\exp(X)$, and the couple $(X, \mathcal{R})$ will be referred to as a *proper relational structure*. Further, given $a \in X$, let us say that the sequence $(x_n; n \geq 0)$ in X is $(a; \mathcal{R})$-*iterative*, provided

$$x_0 = a, \text{ and } x_n \mathcal{R} x_{n+1} \text{ (i.e. } x_{n+1} \in \mathcal{R}(x_n)) \quad \text{for all } n.$$

Proposition 1. *Let the relational structure $(X, \mathcal{R})$ be proper. Then, for each $a \in X$, there is at least an $(a; \mathcal{R})$-iterative sequence in X.*

This principle — proposed, independently, by Bernays [5] and Tarski [26] — is deductible from (AC), but not conversely, cf. Wolk [33]. Moreover, by the developments in Moskhovakis [17, Chapter 8], and Schechter [25, Chapter 6], the *reduced system* (ZF−AC+DC) is comprehensive enough so as to cover the "usual" mathematics; see also Moore [16, Appendix 2].

A basic consequence of (DC) is the so-called *Denumerable Axiom of Choice* (AC(N)).

Proposition 2. *Let $F : N \to \exp(X)$ be a function. Then, for each $a \in F(0)$, there exists a function $f : N \to X$ with $f(0) = a$ and $[f(n) \in F(n), \forall n]$.*

Proof. Denote $Q = N \times X$, and let us introduce the (proper) relation $\mathcal{R}$ over it according to

$$\mathcal{R}(n, x) = \{n + 1\} \times F(n + 1), \quad n \in N, \, x \in X.$$

By an application of (DC) to the proper relational structure $(Q, \mathcal{R})$, the conclusion follows; we do not give details. $\quad\square$

As a consequence of the above facts,

(DC) $\Longrightarrow$ (AC(N)) in the *strongly reduced system* (ZF-AC), whence
(AC(N)) is deductible in the *reduced system* (ZF$-$AC+DC).

The reciprocal of this inclusion is not true; see Moskhovakis [17, Chapter 8, Section 8.25], for details.

3. Matkowski Admissible Functions

Define, for each sequence $(r_n; n \geq 0)$ in R and each point $r \in R$,

$\lim_n r_n = r+$ (also written as $r_n \to r+$) if $\lim_n r_n = r$ and $(r_n > r, \forall n)$.

Denote by $\mathcal{F}(\mathrm{re})(R_+^0, R)$ the family of all *regressive* $\varphi \in \mathcal{F}(R_+^0, R)$ [in the sense: $\varphi(t) < t, \forall t \in R_+^0$]. For each $\varphi \in \mathcal{F}(\mathrm{re})(R_+^0, R)$, define the sequential properties

(M-a) φ is *Matkowski admissible*:
for each $(t_n; n \geq 0)$ in R_+^0 with $(t_{n+1} \leq \varphi(t_n), \forall n)$, we have $\lim_n t_n = 0$.
(str-M-a) φ is *strongly Matkowski admissible*:
for each $(t_n; n \geq 0)$ in R_+^0 with $(t_{n+1} \leq \varphi(t_n), \forall n)$, we have $\sum_n t_n < \infty$.

(The conventions (M-a) and (str-M-a) are taken from Matkowski [12,13] and Turinici [29], respectively). Clearly, for each $\varphi \in \mathcal{F}(\mathrm{re})(R_+^0, R)$,

strongly Matkowski admissible implies Matkowski admissible;

but the converse is not in general true.

(A) A concrete circumstance under which the former of these properties hold may be described as follows. Given $\varphi \in \mathcal{F}(\mathrm{re})(R_+^0, R)$, define the property

(MK-a) φ is *Meir–Keeler admissible*:

$\forall \varepsilon > 0, \exists \delta > 0$, such that $(\varepsilon < s < \varepsilon + \delta)$ implies $\varphi(s) \leq \varepsilon$.

(This convention is suggested by the developments in Meir and Keeler [15]).

Theorem 2. *For each* $\varphi \in \mathcal{F}(\mathrm{re})(R^0_+, R)$, *we have in* $(ZF{-}AC{+}DC)$

$$(M - a) \implies (MK - a) \implies (M - a).$$

Hence, these properties are equivalent over $\mathcal{F}(\mathrm{re})(R^0_+, R)$.

For a complete proof, see Turinici [32, Section 13].

To get concrete examples of such functions, the following constructions are in effect. Let $\varphi \in \mathcal{F}(R^0_+, R)$ be given. Denote, for each $s > 0$,

$$\Lambda^+\varphi(s) \;=\; \inf_{\varepsilon>0} \Phi^*(s+)(\varepsilon), \text{ where } (\Phi^*(s+)(\varepsilon) \;=\; \sup \varphi]s, s + \varepsilon[; \varepsilon > 0),$$
$$\Lambda_+\varphi(s) \;=\; \sup_{\varepsilon>0} \Phi_*(s+)(\varepsilon), \text{ where } (\Phi_*(s+)(\varepsilon) \;=\; \inf \varphi]s, s + \varepsilon[; \varepsilon > 0);$$

these will be referred to as *right superior/inferior limit* of φ at s. Further, given $\varphi \in \mathcal{F}(\mathrm{re})(R^0_+, R)$, define the property

(BW-adm) φ *is Boyd–Wong admissible:* $\Lambda^+\varphi(s) < s$, *for all* $s > 0$.

(This convention is related to the developments in Boyd and Wong [6]). In particular, $\varphi \in \mathcal{F}(\mathrm{re})(R^0_+, R)$ is Boyd–Wong admissible, provided

φ is upper semicontinuous at the right on R^0_+: $\Lambda^+\varphi(s) \leq \varphi(s)$, $\forall \, s \in R^0_+$.

This, for example, is fulfilled when φ is continuous at the right on R^0_+; for, in such a case,

$$\Lambda^+\varphi(s) = \varphi(s), \quad \text{for each } s \in R^0_+.$$

The following auxiliary fact will be useful.

Proposition 3. *Let* $\varphi \in \mathcal{F}(\mathrm{re})(R^0_+, R)$ *be Boyd–Wong admissible. Then,* φ *is Meir–Keeler admissible (or, equivalently: Matkowski admissible).*

Proof. Suppose that $\varphi \in \mathcal{F}(\mathrm{re})(R^0_+, R)$ is Boyd–Wong admissible, and fix $\gamma > 0$; hence, $\Lambda^+\varphi(\gamma) < \gamma$. By definition, there exists $\beta = \beta(\gamma) > 0$ such that $\gamma < t < \gamma + \beta$ implies $\varphi(t) < \gamma$; and this gives the desired fact. $\square$

(B) Passing to the strong Matkowski admissible property, the basic criteria with a practical meaning are the ones below.

Call $h \in \mathcal{F}(R_+^0)$, γ-locally int-normal (where $\gamma > 0$), provided

(lo-in-1) $h(\cdot)$ is decreasing on $]0, \gamma[$

(lo-in-2) $H(t) := \int_0^t h(\xi)d\xi < \infty$, for each $t \in]0, \gamma[$.

We stress that, by the former condition (lo-in-1),

$\int_0^t h(\xi)d\xi := \lim_{s \to 0+} \int_s^t h(\xi)d\xi$ exists in $R_+ \cup \{\infty\}$ for each $t \in]0, \gamma[$;

so, the latter condition (lo-in-2) is meaningful. Moreover, by these definitions,

(p1) $H(\cdot)$ is strictly increasing on $]0, \gamma[$ ($t_1 < t_2 \implies H(t_1) < H(t_2)$);

(p2) $H(\cdot)$ is continuous on $]0, \gamma[$, and $0 = H(0 + 0) := \lim_{t \to 0+} H(t)$.

Note at this moment that the invariance property holds

h is γ-locally int-normal implies $ah + b$ is γ-locally int-normal, $\forall \, a, b > 0$.

When $\gamma > 0$ is generic here, we then say that $h \in \mathcal{F}(R_+^0)$ is *locally int-normal*. As before, the invariance property holds

h is locally int-normal implies $ah + b$ is locally int-normal, $\forall \, a, b > 0$.

Given $\varphi \in \mathcal{F}(re)(R_+^0, R)$, let us associate the function $g \in \mathcal{F}(R_+^0)$, according to

$$(g(t) = t/(t - \varphi(t)); t > 0); \text{ in short: } g = I/(I - \varphi),$$

where $(I(t) = t; t > 0)$ is the *identical function* of $\mathcal{F}(R_+^0)$. Further, call $g \in \mathcal{F}(R_+^0)$, γ-*locally int-subnormal* (where $\gamma > 0$) provided

$g(t) \le h(t)$, $\forall \, t \in]0, \gamma[$, where $h \in \mathcal{F}(R_+^0)$ is γ-locally int-normal.

When $\gamma > 0$ is generic here, we then say that $g \in \mathcal{F}(R_+^0)$ is *locally int-subnormal*.

The following locally int-subnormal-type strongly Matkowski criterion (str-M-lis) is available.

Theorem 3. *Let the function $\varphi \in \mathcal{F}(\mathrm{re})(R_+^0, R)$ be such that its associated function $g = I/(I - \varphi)$ is locally int-subnormal. In addition, suppose that one of the following conditions holds:*

(32-i) φ *is Matkowski admissible (or, equivalently: Meir–Keeler admissible);*

(32-ii) φ *is Boyd–Wong admissible.*

Then, φ is strongly Matkowski admissible (see above).

Proof. There are two steps to be passed.

(i) By the imposed condition, there exists $\gamma > 0$ and $h \in \mathcal{F}(R_+^0)$ with

$$g(t) \le h(t), \ \forall \, t \in]0, \gamma[, \text{ and } h \text{ is } \gamma\text{-locally int-normal.}$$

Moreover, denoting by H its associated primitive, we have (see above)

$$H \text{ is strictly increasing, continuous on }]0, \gamma[\text{ and } H(0+0) = 0.$$

Let the sequence $(t_n; n \ge 0)$ in R_+^0 be such that

(iter) $t_{n+1} \le \varphi(t_n)$, for all n; clearly, (t_n) is strictly descending.

By the Matkowski admissible condition, $t_n \to 0$ as $n \to \infty$, whence

$$\text{there exists } m = m(\gamma) \in N, \text{ such that } t_n < \gamma, \ \forall \, n \ge m.$$

Having these precise, let $i \ge m$ be arbitrary fixed. By the above choice,

$$t_i - \varphi(t_i) \le t_i - t_{i+1}, \text{ whence } 1 \le (t_i - t_{i+1})/(t_i - \varphi(t_i)).$$

Combining with the decreasing property of $h(\cdot)$ yields (by the definition of $H(\cdot)$)

$$t_i \le (t_i - t_{i+1})g(t_i) \le (t_i - t_{i+1})h(t_i) \le H(t_i) - H(t_{i+1}).$$

This gives us (by means of $t_i \to 0$ as $i \to \infty$)

$\sum_{n \ge m} t_n \le H(t_m) - H(0+0) = H(t_m) < \infty$; hence the conclusion.

(ii) Evident in view of every Boyd–Wong admissible function in $\mathcal{F}(\mathrm{re})(R_+^0, R)$ being Matkowski admissible. $\square$

(C) In the following, a basic particular case of this result is discussed. Given $\varphi \in \mathcal{F}(\mathrm{re})(R_+^0, R)$, let us introduce the property

$$(z - con) \quad \varphi \text{ is } zero - contractive: \limsup_{t \to 0+}[\varphi(t)/t] < 1.$$

The following zero-contractive-type strongly Matkowski admissible criterion (str-M-zc) is to be stated.

Theorem 4. *Let the function* $\varphi \in \mathcal{F}(\mathrm{re})(R_+^0, R)$ *be zero contractive. In addition, suppose that one of the following conditions holds:*

(33-i) φ *is Matkowski admissible (or, equivalently: Meir–Keeler admissible);*

(33-ii) φ *is Boyd–Wong admissible.*

Then, φ *is strongly Matkowski admissible.*

Proof. By the zero-contractive property, there exist $\lambda \in]0,1[$, $\gamma > 0$, such that

(rela) $\varphi(t) \leq \lambda t$ for all $t \in]0, \gamma[$.

From this point on, we have two ways of concluding the argument.
    (Standard) (i) Let the sequence $(t_n; n \geq 0)$ in R_+^0 be such that

$t_{n+1} \leq \varphi(t_n)$ for all $n \geq 0$; hence, $(t_n; n \geq 0)$ is strictly descending.

From the Matkowski admissible condition, $t_n \to 0$ as $n \to \infty$; so, there must be some index $m = m(\gamma)$, such that

$(\forall n \geq m): 0 < t_n < \gamma$, whence [by (rela)] $\varphi(t_n) \leq \lambda t_n$.

Combining with the inductive property of our sequence, one gets

$t_{n+1} \leq \lambda t_n$ for all $n \geq m$, wherefrom $\sum_n t_n < \infty$.

(ii) Evident in view of every Boyd–Wong admissible function in $\mathcal{F}(\mathrm{re})(R_+^0, R)$ being Matkowski admissible.
    (Relative) By (rela), we have for the associated function $g := I/(I - \varphi)$

$g(t) := t/(t - \varphi(t)) \leq 1/(1 - \lambda), \quad t \in]0, \gamma[,$

whence g is γ-locally int-subnormal. Combining with the locally int-subnormal-type strongly Matkowski admissible criterion (str-M-lis) gives our conclusion. $\qquad\square$

A limiting version of this statement is to be obtained under the following lines. Call the function $h \in \mathcal{F}(R_+^0)$, *lim-right normal* provided

the right limit $h(s + 0) := \lim_{t \to s+} h(t)$ exists
as an element of R_+ for each $s \in R_+$.

Note at this moment that the invariance property holds:

h is lim-right-normal implies $ah+b$ is lim-right-normal, $\forall\, a, b > 0$.

The following particular version of this property is to be noted. Given $h \in \mathcal{F}(R_+^0)$, call it *monotonic-normal*, when

(norm) h is either increasing or (decreasing, bounded).

As before, the invariance property holds:

h is monotonic-normal implies $ah + b$ is monotonic-normal, $\forall\, a, b > 0$.

It is now clear that

for each $h \in \mathcal{F}(R_+^0)$: h is monotonic-normal implies h is lim-right-normal.

Remark 1. The converse of this functional inclusion is not in general true. In fact, take $h \in \mathcal{F}(R_+^0)$ as

for each $(\forall n \in N)$: $h(t) = t - n$, if $n < t \leq n+1$.

Clearly, h is lim-right-normal. In addition, h is not monotonic-normal because

$(\forall n \geq 1)\ h(n) = 1 > 0 = h(n + 0)$, whence h is not increasing;
$(\forall n \geq 0)\ h$ is increasing on $]n, n+1]$, whence h is not decreasing,

and this proves our claim.

Given $\varphi \in \mathcal{F}(re)(R_+^0, R)$, let us associate the function $g \in \mathcal{F}(R_+^0)$, according to

$(g(t) = t/(t - \varphi(t)); t > 0)$; in short: $g = I/(I - \varphi)$.

Further, letting $g \in \mathcal{F}(R_+^0)$, define the couple of properties:

(l-r-sub) g is *lim-right-subnormal*, provided
$g(t) \le h(t)$, $t \in R_+^0$, where $h \in \mathcal{F}(R_+^0)$ is lim-right-normal;
(adm-sub) g is *monotonic-subnormal*, provided
$g(t) \le h(t)$, $t \in R_+^0$, where $h \in \mathcal{F}(R_+^0)$ is monotonic-normal.

By the preceding observation, we have that

for each $g \in \mathcal{F}(R_+^0)$: g is monotonic-subnormal implies
g is lim-right-subnormal.

Putting these together, one derives the following practical strongly Matkowski admissible criterion.

Theorem 5. *Let the function $\varphi \in \mathcal{F}(\mathrm{re})(R_+^0, R)$ be such that one of the following extra conditions holds:*

(ec-1) *the associated function $g = I/(I - \varphi)$ is lim-right-subnormal;*
(ec-2) *the associated function $g = I/(I - \varphi)$ is monotonic-subnormal.*

Then, φ is strongly Matkowski admissible.

Proof. It will suffice to verify the first part. By the imposed condition,

$$g(t) \le h(t) \le k(t) := h(t) + 2, \quad t \in R_+^0$$

for some function $h \in \mathcal{F}(R_+^0)$ with $h = $ lim-right-normal (hence, so is $k := h + 2$). From this relation, we derive

$(\forall t \in R_+^0)$: $\varphi(t) \le t(k(t) - 1)/k(t)$; hence, $\varphi(t)/t \le (k(t) - 1)/k(t)$.

Passing to superior limit as $t \to s+$ (where $s > 0$ is arbitrary fixed) yields

$$\limsup\nolimits_{t \to s+} \varphi(t) \le s(k(s + 0) - 1)/k(s + 0) < s, \quad \forall s > 0,$$

which tells us that φ is Boyd-Wong admissible; hence, Matkowski admissible. On the other hand, passing to superior limit as $t \to 0+$ yields (via $2 \le k(0 + 0) < \infty$)

$\limsup\nolimits_{t \to 0+} [\varphi(t)/t] \le (k(0 + 0) - 1)/k(0 + 0) < 1$; so, $\varphi = $ zero contractive.

Summing up, the preceding statement is applicable here, and we are done.   □

(D) A global version of the (local) strongly Matkowski admissible result above may be stated as follows. Call the function $h \in \mathcal{F}(R_+^0)$, *int-normal*, provided

(i-n-1) $h(\cdot)$ is decreasing on R_+^0;
(i-n-2) $H(t) := \int_0^t h(\xi)d\xi < \infty$ for each $t > 0$.

Note that, by the former condition (i-n-1),

$\int_0^t h(\xi)d\xi := \lim_{s\to 0+} \int_s^t h(\xi)d\xi$ exists in $R_+ \cup \{\infty\}$, for each $t > 0$;

so, the latter condition (i-n-2) is meaningful. Moreover, by these definitions,

(p-1) $H(\cdot)$ is strictly increasing on R_+^0 ($t_1 < t_2 \implies H(t_1) < H(t_2)$);
(p-2) $H(\cdot)$ is continuous on R_+^0 and $0 = H(0+0) := \lim_{t\to 0+} H(t)$.

In addition, we have that the invariance property holds:

h is int-normal implies $ah + b$ is int-normal for each $a, b > 0$.

Given $\varphi \in \mathcal{F}(re)(R_+^0, R)$, let us associate it the function $g \in \mathcal{F}(R_+^0)$, according to

$(g(t) = t/(t - \varphi(t)); t > 0)$; in short: $g = I/(I - \varphi)$.

Further, call $g \in \mathcal{F}(R_+^0)$, *int-subnormal*, provided

$g(t) \le h(t)$, $t \in R_+^0$, where $h \in \mathcal{F}(R_+^0)$ is int-normal.

The following int-subnormal-type strongly Matkowski criterion (str-M-is) is available.

Theorem 6. *Let the function* $\varphi \in \mathcal{F}(re)(R_+^0, R)$ *be such that*

the associated function $g = I/(I - \varphi)$ *is int-subnormal.*

Then, φ is strongly Matkowski admissible (see above).

Proof. By the imposed condition, we have, for $g = I/(I - \varphi)$,

$g(t) \le h(t) \le k(t) := h(t) + 2$, $t \in R_+^0$, for some $h \in \mathcal{F}(R_+^0)$ with h=int-normal; hence, so is $k := h + 2$.

This yields in a direct mode

$(\forall t \in R_+^0)$: $\varphi(t) \le t(k(t) - 1)/k(t)$; hence, $\varphi(t)/t \le (k(t) - 1)/k(t)$.

Passing to superior limit as $t \to s+$ (where $s > 0$ is arbitrary fixed) gives

$$\limsup_{t \to s+} \varphi(t) \le s(k(s+0) - 1)/k(s+0) < s, \quad \forall s > 0,$$

which tells us that φ is Boyd-Wong admissible; hence, Matkowski admissible.

From this point on, we have two ways of concluding the argument.

(Standard) Let the sequence $(t_n; n \ge 0)$ in R_+^0 be such that

$$t_{n+1} \le \varphi(t_n) \text{ for all } n; \text{ hence, } (t_n; n \ge 0) \text{ is strictly descending.}$$

By the Matkowski admissible condition, $t_n \to 0$ as $n \to \infty$. On the other hand, for the arbitrary fixed $i \ge 0$, we have (by the above choice)

$$t_i - \varphi(t_i) \le t_i - t_{i+1}, \text{ whence } 1 \le (t_i - t_{i+1})/(t_i - \varphi(t_i)).$$

Combining with the decreasing property of $h(\cdot)$ yields (by the definition of $H(\cdot)$)

$$t_i \le (t_i - t_{i+1})g(t_i) \le (t_i - t_{i+1})h(t_i) \le H(t_i) - H(t_{i+1}), \forall i.$$

This yields (in view of $t_i \to 0$ as $i \to \infty$)

$$\sum_n t_n \le H(t_0) - H(0+0) = H(t_0) < \infty; \text{ hence the conclusion.}$$

(Relative) Clearly, the following is valid:

$(\forall g \in \mathcal{F}(R_+^0))$: g is int-subnormal implies g is locally int-subnormal.

This, along with the obtained Boyd–Wong admissible property tells us that the locally int-subnormal-type strongly Matkowski admissible criterion (str-M-lis) is applicable here, and we are done.   □

In particular, if g is int-normal, the obtained criterion is nothing else than the 1975 one due to Altman [2].

Summing up, we derived three strongly Matkowski admissible criteria:

(cr-1) the locally int-subnormal-type strongly Matkowski admissible criterion (str-M-lis);

(cr-2) the zero-contractive-type strongly Matkowski admissible criterion (str-M-zc);

(cr-3) the int-subnormal-type strongly Matkowski admissible criterion (str-M-is).

For the moment, we have established that

(inclu) (str-M-lis) $\Longrightarrow$ (str-M-zc), (str-M-lis) $\Longrightarrow$ (str-M-is),

which tells us that (str-M-lis) is the best strong Matkowski criterion in this series. Concerning the remaining criteria (str-M-is) and (str-M-zc), a natural question is to establish which criterion is the next one in this preference order. As expected, the criterion in question is (str-M-is). The following example will certify this.

Example 1. Let the function $h \in \mathcal{F}(R_+^0)$ be taken according to

(i) h is int-normal (hence: locally int-normal);
(ii) $(h(t) \geq 2, \forall t \in R_+^0)$, and $h(0+0) = \infty$.

[Note that, for each $\nu \in]0,1[$, the function $(h(t) = 2 + t^{-\nu}; t \in R_+^0)$ fulfills all these conditions; hence, the class of all such objects is pretty large.] Then, let the function $\varphi \in \mathcal{F}(R_+^0)$ be introduced as

$$\varphi(t) = t(h(t) - 1)/h(t) = t[1 - 1/h(t)], \quad t \in R_+^0.$$

By this very convention,

(pro-1) the attached function $g := I/(I - \varphi)$ fulfills $g = h$, so that g is int-subnormal (hence: locally int-subnormal).

Further, φ is Boyd–Wong admissible in view of

(pro-2) $\Lambda^+\varphi(s) = \varphi(s+0) = s[h(s+0) - 1]/h(s+0) < s, \forall s \in R_+^0$.

Consequently, both locally int-subnormal-type strongly Matkowski admissible criterion (str-M-lis) and int-subnormal-type strongly Matkowski criterion (str-M-is) are applicable in our context, wherefrom φ is strongly Matkowski admissible. On the other hand, since (cf. the assumptions upon h)

(pro-3) $\lim_{t \to 0+}[\varphi(t)/t] = \lim_{t \to 0+}[1 - 1/h(t)] = 1$,

the zero-contractive-type strongly Matkowski admissible criterion (str-M-zc) is not applicable here. Putting these together yields the desired conclusion.

As precise, this is only a partial answer to the posed problem. Further aspects will be delineated elsewhere.

4. Main Result

Let (X, d) be a metric space. Given the self-map $T \in \mathcal{F}(X)$, we have to determine whether $\text{Fix}(T)$ is nonempty; and, if this holds, to establish whether T is fix-asingleton [or, equivalently, T is fix-singleton]. The specific directions under which this problem is to be solved were already listed. Sufficient conditions for getting such properties are being founded on the *orbital full* concepts (in short: (o-f)-concepts). Namely, call the sequence $(z_n; n \geq 0)$ in X,

(i) *T-orbital* when $(z_n = T^n x; n \geq 0)$ for some $x \in X$,
(ii) *full* if $n \mapsto z_n$ is injective $(z_i \neq z_j$ for $i \neq j)$;

the intersection of these notions is just the precise one.

    (reg-1) Call X, $(o\text{-}f,d)$-*complete*, provided (for each (o-f)-sequence) d-Cauchy $\Longrightarrow$ d-convergent.

    (reg-2) We say that T is $(o\text{-}f,d)$-*continuous* if $[(z_n)=(o\text{-}f)$-sequence and $z_n \xrightarrow{d} z]$ imply $Tz_n \xrightarrow{d} Tz$.

    Finally, when the orbital properties are ignored, these conventions may be written in the usual way; we do not give details.

    As a basic completion of this, we have to discuss the contractive-type conditions to be used. Denote, for each $x, y \in X$,

$$A_1(x, y) = d(x, y), \quad A_2(x, y) = \max\{d(x, Tx), d(y, Ty)\},$$
$$A_3(x, y) = (1/2)[d(x, Ty) + d(Tx, y)], \quad \mathcal{A} = \{A_1, A_2, A_3\}.$$

By taking all possible maxima between these, one gets $2^3 - 1 = 7$ functions of this type (including the ones we just listed); these may be written as

$$\max(\mathcal{G}), \text{ where } \mathcal{G} \in \exp(\mathcal{A}).$$

Having these precise, let $\psi : R_+^0 \to R$ be a function. For simplicity, we will denote by $\varphi : R_+^0 \to R$ its associated function

$$(\varphi(t) = t\psi(t); t \in R_+^0); \text{ in short: } \varphi = I\psi.$$

Further, let $P : X \times X \to R_+$ be a map. We say that T is (ψ, P)-*contractive* if

$$\min\{d(Tx, Ty), d(y, Ty)\} \le \psi(d(x, y))P(x, y) \text{ for all } x, y \in X,$$
$$x \ne y.$$

The properties of φ to be used here are taken from the ones we already listed. Concerning the properties of the mapping P, the following ones are of interest.

(I) Let us say that P is (d, T)-*orbitally-bounded*, provided

(o-bd) $P(x, Tx) \le A_2(x, Tx)$, whenever $x \ne Tx$.

For example, any mapping $P = A_i$, where $i \in \{1, 3\}$ fulfills such a condition; hence, all the more, any mapping $P = \max(\mathcal{G})$, where $\mathcal{G} \in \exp\{A_1, A_3\}$.

(II) Let us introduce the property

(d-asy-sing) P is (d, T)-*singular-asymptotic*:
for each (o-f)-sequence (x_n) in X, and each $z \in X$ with
$(x_n \xrightarrow{d} z, d(z, Tz) > 0)$, we must have $\liminf_n P(x_n, z) < d(z, Tz)$.

Clearly, the following property is a particular case of this:

(str-d-asy-sing) P is *strongly* (d, T)-*singular-asymptotic*:
for each (o-f)-sequence (x_n) in X, and each $z \in X$ with
$(x_n \xrightarrow{d} z, d(z, Tz) > 0)$, we must have $(\exists) \lim_n P(x_n, z) < d(z, Tz)$.

For example, any mapping $P = A_i$, where $i \in \{1, 3\}$ is strongly (d, T)-singular-asymptotic; hence, all the more, any mapping $P = \max(\mathcal{G})$, where $\mathcal{G} \in \exp\{A_1, A_3\}$.

We are now in position to state the main result of this exposition.

Theorem 7. *Suppose that T is (ψ, P)-contractive, where $\psi : R_+^0 \to R$ and $P : X \times X \to R_+$ fulfill the following:*

(41-i) *the associated to ψ function $\varphi = I\psi$ is regressive and strongly Matkowski admissible;*

(41-ii) *P is (d,T)-orbitally-bounded and (d,T)-singular-asymptotic.*

In addition, let X be $(o\text{-}f, d)$-complete. Then, T is strong Picard (modulo d).

Proof. Let $x_0 \in X$ be arbitrary fixed, and put $(x_n = T^n x_0; n \geq 0)$; hence (x_n) is an orbital sequence. The alternative

(x_n) is telescopic: $x_h = x_{h+1}$, for some $h \geq 0$,

ends the argument because $x_n \xrightarrow{d} x_h$ and $x_h \in \mathrm{Fix}(T)$. Suppose now that

(x_n) is not telescopic: $\rho_n := d(x_n, x_{n+1}) > 0$ for all n.

There are two steps to be passed.

Step 1. Let $n \geq 0$ be arbitrary fixed. By the contractive condition,

(iter-1) $\rho_{n+1} \leq \psi(\rho_n) P(x_n, x_{n+1})$; hence, in particular, $\psi(\rho_n) > 0$.

From the regressiveness of $\varphi = I\psi$, we have on the one hand

(rela) $\rho_n \psi(\rho_n) = \varphi(\rho_n) < \rho_n$; hence, $\psi(\rho_n) < 1$.

Moreover, by the orbital boundedness property, we have on the other hand

$$P(x_n, x_{n+1}) \leq A_2(x_n, x_{n+1}) = \max\{\rho_n, \rho_{n+1}\}.$$

Replacing these in (iter-1), one derives

(iter-2) $(\forall n)$: $\rho_{n+1} \leq \psi(\rho_n) \max\{\rho_n, \rho_{n+1}\}$.

Two basic consequences of this fact may be derived.

(Cons-1) A first consequence of (iter-2) is [via (rela)]

(s-desc) $(\forall n)$: $\rho_{n+1} < \max\{\rho_n, \rho_{n+1}\}$; whence $\rho_{n+1} < \rho_n$,

proving that (ρ_n) is strictly descending in R_+. But then, necessarily,

(x_n) is full: $i \neq j$ implies $x_i \neq x_j$ (whence, $d(x_i, x_j) > 0$).

In fact, suppose by contradiction that

there exists $i, j \in N$ with $i < j$, $x_i = x_j$.

Then, by definition,

$x_{i+1} = x_{j+1}$, so that $\rho_i = \rho_j$,

in contradiction with $\rho_i > \rho_j$, and the assertion follows.
(Cons-2) A second consequence of (iter-2) is

(iter-3) $(\forall n)$: $\rho_{n+1} \le \psi(\rho_n)\rho_n = \varphi(\rho_n)$.

Combining with the strong Matkowski admissible property of φ gives

$\sum_n \rho_n < \infty$; wherefrom, (x_n) is a d-Cauchy sequence.

Step 2. As X is (o-f,d)-complete, $x_n \xrightarrow{d} z$, for some (uniquely determined) $z \in X$. By the full property of $(Tx_n = x_{n+1}; n \ge 0)$,

$E := \{n \in N; Tx_n = Tz\}$ is an asingleton,

so that the following separation property holds:

(sepa) there exists $i = i(z) \ge 0$, such that
$n \ge i$ implies $d(Tx_n, Tz) > 0$ [hence $d(x_n, z) > 0$].

Denote for simplicity $(u_n = x_{n+i}; n \ge 0)$; clearly,

(u_n) is an orbital sequence with $d(u_n, z) > 0$, $d(Tu_n, Tz) > 0$, $\forall n$.

Suppose by contradiction that $b := d(z, Tz) > 0$. We show that this is not compatible with the imposed requirements upon P. In fact, the contractive condition is applicable to the couples $((u_n, z); n \ge 0)$ and yields

(contr) $\min\{d(Tu_n, Tz), b\} \le P(u_n, z)$, $\forall n$.

Passing to lim inf as $n \to \infty$, one gets (by the singular asymptotic hypothesis)

$b = \lim_n \min\{d(Tu_n, Tz), b\} = \liminf_n \min\{d(Tu_n, Tz), b\}$
$\le \liminf_n P(u_n, z) < b$; a contradiction.

Hence, necessarily, $b = 0$ (i.e. $z \in \text{Fix}(T)$), and the proof is complete.

□

Note that an extension of the obtained result to quasi-ordered metric spaces is directly available under the lines in Turinici [30]. This must be combined with the fact that an extended quasi-metric setting of these developments is possible as well; see in this direction Hitzler [10, Chapter 1, Section 1.2], and Turinici [31]. Finally, multivalued versions of Theorem 7 are accessible by following the methods in Nadler [18]. Further aspects will be delineated elsewhere.

5.   Particular Case

In the following, some basic particular cases of our main result are discussed. These, as expected, involve the function $\psi : R_+^0 \to R$ and the mapping $P : X \times X \to R_+$ appearing there.

Part-Case 1. Let $\psi : R_+^0 \to [0,1[$ be a function. The following condition will be used here:

(u-BW) ψ is *unitary Boyd–Wong admissible*:
$\limsup_{s \to t+} \psi(s) < 1$, for all $t > 0$.

Let $\varphi = I\psi$ be the associated to ψ function. Clearly, φ is regressive. Moreover, by elementary calculations, we have

φ is Boyd–Wong admissible: $\limsup_{s \to t+} \varphi(s) < t$, for all $t > 0$.

According to some previous results, we have that, whenever one of the extra conditions holds,

(ec-1) $g := I/(I - \varphi) = 1/(1 - \psi)$ is locally int-subnormal,
(ec-2) $\limsup_{t \to 0+}[\varphi(t)/t] = \limsup_{t \to 0+} \psi(t) < 1$
(referred to as ψ is *zero-subunitary*),

the associated function φ is strongly Matkowski admissible.
    Putting these together, one derives the following fixed point statement with a methodological importance.

Theorem 8. *Let the self-map $T \in \mathcal{F}(X)$ be (ψ, P)-contractive, where the unitary Boyd–Wong admissible function $\psi : R_+^0 \to [0,1[$ fulfills (in addition)*

either $g := 1/(1 - \psi)$ is locally int-subnormal or ψ is zero-subunitary *and the mapping $P : X \times X \to R_+$ is (d, T)-orbitally-bounded and (d, T)-singular-asymptotic. In addition, let X be $(o\text{-}f, d)$-complete. Then, necessarily, T is strong Picard (modulo d).*

Part-Case 2. Let (π) be a certain property over $\mathcal{F}(X \times X, R_+)$; we say that it is max-*invariant*, provided

> if the mappings $P_1, P_2 \in \mathcal{F}(X \times X, R_+)$ have the property (π), then
> $P := \max\{P_1, P_2\}$ has the property (π) as well.

We have to establish whether this max invariance is retainable for the orbital boundedness and the singular asymptotic properties over $\mathcal{F}(X \times X, R_+)$. Concerning the (d, T)-orbitally-bounded property, the following positive answer is valid.

Proposition 4. *The (d, T)-orbitally-bounded property is max-invariant.*

Proof. Suppose that $P_1 : X \times X \to R_+$ and $P_2 : X \times X \to R_+$ are (d, T)-orbitally-bounded. By definition,

$$P_1(x, Tx) \leq A_2(x, Tx), \; P_2(x, Tx) \leq A_2(x, Tx), \quad \forall x \in X, \, x \neq Tx.$$

But then, evidently,

$$P(x, Tx) = \max\{P_1(x, Tx), P_2(x, Tx)\} \leq A_2(x, Tx), \quad \forall x \in X, \\ x \neq Tx,$$

which tells us that $P = \max\{P_1, P_2\}$ is (d, T)-orbitally-bounded. $\square$

Concerning the (d, T)-singular-asymptotic property, the max invariance of it seems to be not available in general, as the following example shows.

Example 2. Let the sequence (x_n) in X and the point $z \in X$ be such that $[x_n \xrightarrow{d} z, \, 0 < d(z, Tz) < 1]$. Then, take the mappings $P_1, P_2 : X \times X \to R_+$ as

$$P_1(x_{2k}, z) = 0, \, P_1(x_{2k+1}, z) = 1, \, k \in N;$$
$$P_2(x_{2k}, z) = 1, \, P_2(x_{2k+1}, z) = 0, \, k \in N;$$
$$\text{hence, } (P(x_n, z) = 1, \, n \in N), \text{ where } P = \max\{P_1, P_2\}.$$

But then, the following evaluations hold:

$$\liminf_n P_1(x_n, z) = \liminf_n P_2(x_n, z) = 0 < d(z, Tz);$$
$$\liminf_n P(x_n, z) = \lim_n P(x_n, z) = 1 > d(z, Tz).$$

This, essentially, proves our claim.

As a consequence, the only chance that we have is to establish the underlying invariance for the strongly (d, T)-singular-asymptotic property over the class $\mathcal{F}(X \times X, R_+)$. In this direction, we have the following positive answer.

Proposition 5. *The strongly (d, T)-singular-asymptotic property over the class $\mathcal{F}(X \times X, R_+)$ is* max-*invariant.*

Proof.   Suppose that $P_1 : X \times X \to R_+$ and $P_2 : X \times X \to R_+$ are strongly (d, T)-singular-asymptotic. Letting the (o-f)-sequence (x_n) in X and the point $z \in X$ be such that $[x_n \xrightarrow{d} z, d(z, Tz) > 0]$, we have

$$(\exists) \lim_n P_1(x_n, z) < d(z, Tz) \text{ and } (\exists) \lim_n P_2(x_n, z) < d(z, Tz).$$

But then, evidently,

$$\lim_n P(x_n, z) = \max\{\lim_n P_1(x_n, z), \lim_n P_2(x_n, z)\} < d(z, Tz),$$

which tells us that $P = \max\{P_1, P_2\}$ is strongly (d, T)-singular-asymptotic. $\square$

Clearly, the invariance of these properties may be extended to a finite number of such maps. This observation will be useful in treating the following example with a methodological importance. Define the system of maps over $\mathcal{F}(X \times X, R_+)$ as: for each $x, y \in X$,

$$B_1(x, y) = (1/2)[d(x, y) + d(y, Tx)],$$
$$B_2(x, y) = (1/2)[d(x, y) + d(x, Tx)],$$
$$B_3(x, y) = (1/2)[d(x, y) + d(y, Ty)],$$
$$C_1(x, y) = (1/3)[d(x, y) + d(y, Ty) + d(y, Tx)],$$
$$C_2(x, y) = (1/3)[d(x, y) + d(x, Tx) + d(y, Ty)],$$
$$C_3(x, y) = (1/3)[d(x, y) + d(x, Tx) + d(y, Tx)],$$
$$C_4(x, y) = (1/3)[d(x, y) + d(x, Ty) + d(y, Tx)],$$
$$C_5(x, y) = (1/3)[d(x, y) + d(x, Tx) + d(y, Tx)],$$

$$C_6(x,y) = (1/3)[d(x,y) + d(y,Ty) + d(y,Tx)],$$
$$C_7(x,y) = (1/3)[d(x,y) + d(x,Ty) + d(y,Tx)],$$
$$E_1(x,y) = (1/4)[d(x,y) + d(y,Ty) + d(x,Ty) + d(y,Tx)],$$
$$E_2(x,y) = (1/4)[d(x,y) + d(x,Tx) + d(y,Ty) + d(y,Tx)],$$
$$E_3(x,y) = (1/4)[d(x,y) + d(x,Tx) + d(x,Ty) + d(y,Tx)].$$

Then, put

$$P = \max\{A_1, B_1, B_2, B_3, C_1, \ldots, C_7, E_1, E_2, E_3\}.$$

(I) By the triangular inequality, we have for each $x \in X$

$$A_1(x,Tx) = d(x,Tx) \le A_2(x,Tx),$$
$$B_1(x,Tx) = (1/2)d(x,Tx) \le A_2(x,Tx),$$
$$B_2(x,Tx) = d(x,Tx) \le A_2(x,Tx),$$
$$B_3(x,Tx) = (1/2)[d(x,Tx) + d(Tx,T^2x)] \le A_2(x,Tx),$$
$$C_1(x,Tx) = (1/3)[d(x,Tx) + d(Tx,T^2x)] \le A_2(x,Tx),$$
$$C_2(x,Tx) = (1/3)[2d(x,Tx) + d(Tx,T^2x)] \le A_2(x,Tx),$$
$$C_3(x,Tx) = (1/3)[2d(x,Tx)] \le A_2(x,Tx),$$
$$C_4(x,Tx) = (1/3)[d(x,Tx) + d(x,T^2x)]$$
$$\le (1/3)[2d(x,Tx) + d(Tx,T^2x)] \le A_2(x,Tx),$$
$$C_5(x,Tx) = (1/3)[2d(x,Tx)] \le A_2(x,Tx),$$
$$C_6(x,Tx) = (1/3)[d(x,Tx) + d(Tx,T^2x)] \le A_2(x,Tx),$$
$$C_7(x,Tx) = (1/3)[d(x,Tx) + d(x,T^2x)]$$
$$\le (1/3)[2d(x,Tx) + d(Tx,T^2x)] \le A_2(x,Tx),$$
$$E_1(x,Tx) = (1/4)[d(x,Tx) + d(Tx,T^2x) + d(x,T^2x)]$$
$$\le (1/2)[d(x,Tx) + d(Tx,T^2x)] \le A_2(x,Tx),$$
$$E_2(x,Tx) = (1/4)[2d(x,Tx) + d(Tx,T^2x)] \le A_2(x,Tx),$$
$$E_3(x,Tx) = (1/4)[2d(x,Tx) + d(x,T^2x)]$$
$$\le (1/4)[3d(x,Tx) + d(Tx,T^2x)] \le A_2(x,Tx),$$

and this, along with the auxiliary statement above, yields

$$P(x,Tx) \le A_2(x,Tx), \ \forall x \in X;$$ that is: P is (d,T)-orbitally-bounded.

(II) Let the (o-f)-sequence (x_n) in X and the point $z \in X$ be such that

$$x_n \xrightarrow{d} z \text{ (i.e. } d(x_n, z) \to 0 \text{ as } n \to \infty) \text{ and } d(z, Tz) > 0.$$

This gives (by the iterative choice of (x_n))

$$d(x_n, Tx_n) \to 0, \ d(Tx_n, z) \to 0, \quad \text{as } n \to \infty.$$

In addition, by a metrical property of d, we have

$$d(x_n, Tz) \to d(z, Tz), \quad \text{as } n \to \infty.$$

Combining these, we have (by definition)

$\lim_n A_1(x_n, z) = 0, \ \lim_n B_1(x_n, z) = 0, \ \lim_n B_2(x_n, z) = 0,$
$\lim_n B_3(x_n, z) = (1/2)d(z, Tz), \ \lim_n C_1(x_n, z) = (1/3)d(z, Tz),$
$\lim_n C_2(x_n, z) = (1/3)d(z, Tz), \ \lim_n C_3(x_n, z) = 0,$
$\lim_n C_4(x_n, z) = (1/3)d(z, Tz), \ \lim_n C_5(x_n, z) = 0,$
$\lim_n C_6(x_n, z) = (1/3)d(z, Tz), \ \lim_n C_7(x_n, z) = (1/3)d(z, Tz),$
$\lim_n E_1(x_n, z) = (1/2)d(z, Tz), \ \lim_n E_2(x_n, z) = (1/4)d(z, Tz),$
$\lim_n E_3(x_n, z) = (1/4)d(z, Tz),$

and this yields directly

$$(\exists) \lim_n P(x_n, z) = (1/2)d(z, Tz) < d(z, Tz),$$

proving that P is strongly (d, T)-singular-asymptotic.

Now, putting these together, one derives, via Theorem 8, the following fixed point statement.

Theorem 9. *Let the self-map $T \in \mathcal{F}(X)$ be (ψ, P)-contractive, where the unitary Boyd–Wong admissible function $\psi : R_+^0 \to [0, 1[$ fulfills (in addition)*

either $g := 1/(1 - \psi)$ is locally int-subnormal or ψ is zero-subunitary

and the mapping $P : X \times X \to R_+$ is taken as before. In addition, let X be $(o\text{-}f, d)$-complete. Then, T is strong Picard (modulo d).

In particular, when (see above)

the unitary Boyd–Wong admissible function $\psi : R_+^0 \to [0, 1[$ is (exclusively) zero-subunitary, the obtained statement above includes the main result in Du *et al.* [8]. In fact, there are many other examples of mappings P to be handled via these methods; like, e.g., the ones appearing in the chapter by Du and Rassias [9]. Further aspects will be delineated elsewhere.

References

[1] M. Akkouchi (2011). Common fixed points for weakly compatible maps satisfying implicit relations without continuity. *Demonstratio Math.*, **44**, 151–158.

[2] M. Altman (1975). An integral test for series and generalized contractions. *Amer. Math. Monthly*, **82**, 827–829.

[3] S. Banach, Sur les opérations dans les ensembles abstraits et leur application aux équations intégrales. *Fund. Math.*, **3**, 133–181.

[4] V. Berinde and F. Vetro (2012). Common fixed points of mappings satisfying implicit contractive conditions. *Fixed Point Theory Appl.*, **2012**, 105.

[5] P. Bernays (1942). A system of axiomatic set theory: Part III. Infinity and enumerability analysis. *J. Symbolic Logic*, **7** (1942), 65–89.

[6] D. W. Boyd and J. S. W. Wong (1969). On nonlinear contractions. *Proc. Amer. Math. Soc.*, **20**, 458–464.

[7] P. J. Cohen (1966). *Set Theory and the Continuum Hypothesis* (Benjamin, New York).

[8] W.-S. Du, E. Karapinar, and Z. He (2018). Some simultaneous generalizations of well-known fixed point theorems and their applications to fixed point theory. *Mathematics*, **6**, 117.

[9] W.-S. Du and T. M. Rassias (2020). Simultaneous generalizations of known fixed point theorems for a Meir–Keeler type condition with applications. *Int. J. Nonlinear Anal. Appl.*, **11**, 55–66.

[10] P. Hitzler (2001). *Generalized metrics and Topology in Logic Programming Semantics*, PhD Thesis, National University of Ireland, University College Cork.

[11] S. Leader (1979). Fixed points for general contractions in metric spaces. *Math. Jpn.*, **24**, 17–24.

[12] J. Matkowski (1975). Integrable Solutions of Functional Equations, *Dissertiones Mathematicae*, Vol. 127 (Polish Scientific Publishers, Warsaw).

[13] J. Matkowski (1977). Fixed point theorems for mappings with a contractive iterate at a point. *Proc. Amer. Math. Soc.*, **62**, 344–348.

[14] J. Matkowski (1980). Fixed point theorems for contractive mappings in metric spaces. *Časopis Pest. Mat.*, **105**, 341–344.

[15] A. Meir and E. Keeler (1969). A theorem on contraction mappings. *J. Math. Anal. Appl.*, **28**, 326–329.

[16] G. H. Moore (1982). *Zermelo's Axiom of Choice: Its Origin, Development and Influence* (Springer, New York).

[17] Y. Moskhovakis (2006). *Notes on Set Theory* (Springer, New York).

[18] S. B. Nadler Jr. (1969). Multi-valued contraction mappings, *Pacific J. Math.*, **30**, 475–488.

[19] H. K. Nashine, Z. Kadelburg, and P. Kumam (2012). Implicit-relation-type cyclic contractive mappings and applications to integral equations. *Abstr. Appl. Anal.*, **2012**, 386253.

[20] M. Piticari (1982). Successive approximations method and Fixed Point Principle. *Diss. Thesis, Alexandru Ioan Cuza University of Iași* (in Romanian).

[21] V. Popa (1999). Some fixed point theorems for compatible mappings satisfying an implicit relation. *Demonstratio Math.*, **32**, 157–163.

[22] S. Reich (1972). Fixed points of contractive functions. *Boll. Unione Mat. Ital.*, **5**, 26-42.

[23] B. E. Rhoades (1977). A comparison of various definitions of contractive mappings. *Trans. Amer. Math. Soc.*, **226**, 257–290.

[24] I. A. Rus (2001). *Generalized Contractions and Applications* (Cluj University Press, Cluj-Napoca).

[25] E. Schechter (1997). *Handbook of Analysis and its Foundation* (Academic Press, New York).

[26] A. Tarski (1948). Axiomatic and algebraic aspects of two theorems on sums of cardinals. *Fund. Math.*, **35**, 79–104.

[27] M. Turinici (1976). Fixed points of implicit contraction mappings. *An. Ştiinţ. Univ. Al. I. Cuza Iași (Mat.)*, **22**, 177–180.

[28] M. Turinici (1980). Fixed points of implicit contractions via Cantor's intersection theorem. *Bul. Inst. Polit. Iași (Sect I: Mat., Mec. Teor., Fiz.)*, **26**(30), 65–68.

[29] M. Turinici (2013). Wardowski implicit contractions in metric spaces. *arXiv: 1211-3164-v2*.

[30] M. Turinici (2014). Contraction maps in ordered metrical structures. In *Mathematics Without Boundaries: Surveys in Interdisciplinary Research.* P. M. Pardalos and T. M. Rassias, (eds.), (Springer, New York), pp. 533–575.

[31] M. Turinici (2014). Contractive maps in locally transitive relational metric spaces. *Sci. World J.*, **2014**, 169358.

[32] M. Turinici (2020). *Reports in Metrical Fixed Point Theory* (Pim Editorial House, Iași).

[33] E. S. Wolk (1983). On the principle of dependent choices and some forms of Zorn's lemma. *Canad. Math. Bull.*, **26**, 365–367.

Chapter 11

Cryptographic Properties of Boolean Functions Generating Similar de Bruijn Sequences

Andreas Varelias[*,§], Konstantinos Limniotis[*,†,¶,‖],
and Nicholas Kolokotronis[‡,**]

[*]*School of Pure and Applied Sciences, Open University of Cyprus,
Latsia, Nicosia, Cyprus*
[†]*Department of Informatics and Computers Engineering,
University of West Attica, Egaleo, Athens, Greece*
[‡]*Department of Informatics and Telecommunications,
University of Peloponnese, Tripolis, Greece*

[§]*andreas.varelias1@st.ouc.ac.cy;*
[¶]*konstantinos.limniotis@ouc.ac.cy;*
[‖]*klimn@uniwa.gr*
[**]*nkolok@uop.gr*

Boolean functions generating de Bruijn sequences are discussed in this chapter in terms of investigating whether two Boolean functions that generate "similar" de Bruijn sequences may have significant differences in their cryptographic properties. More precisely, having as starting point some recent results on identifying pairs of de Bruijn sequences sharing the longest common subsequence, we present a new approximation algorithm, utilizing the (inverse) suffix arrays of the sequences in order to compute pairs of such de Bruijn sequences. Therefore, having this algorithm as the means to compute similar de Bruijn sequences, we subsequently examine cryptographic properties of the corresponding Boolean functions, namely, the algebraic degree, the nonlinearity and the algebraic immunity; moreover, the corresponding linear complexities of the de Bruijn sequences were also studied. We show that, although for the majority of the cases these properties remain invariant, we may have some differences — especially in the nonlinearity and also in the algebraic immunity — which could be of cryptanalytic value.

1. Introduction

de Bruijn sequences receive much attention from the research community during the last decades, since they play an important role in several application fields. Especially in cryptography, de Bruijn sequences are associated with maximum period Nonlinear Feedback shift registers (NLFSRs) [1], which in turn are present in several modern ciphers see, e.g. Grain [2], being one of the 10 finalists in NIST's competition for standardizing lightweight cryptographic algorithms. In this perspective, there are still several open research questions, such as how to construct a Boolean function which, being the feedback function of an NLFSR, yields a de Bruijn generator (known solutions exist only for some special cases or for small NLFSRs) [3] as well as which are the cryptographic properties of such a Boolean function.

Very recently, the problem of finding, for a given de Bruijn sequence y of order n, another de Bruijn sequence $\tilde{y}$ with the same order having the longest possible common subsequence with y is being investigated [4]. As it is shown therein, such a sequence $\tilde{y}$ can be always obtained by y by an appropriate single cross-join operation, namely, with the property that the corresponding cross-join pairs have the maximum diameter (as described in Ref. [4]). There is no explicit method though for determining a cross-join pair with the maximum diameter.

Although it is not explicitly mentioned in Ref. [4], the above problem is of high cryptographic importance; more precisely, for a given cryptographic sequence, finding another one which resembles the initial one but has cryptographic weaknesses may be the starting point for mounting successful cryptanalytic attacks; for example, linear [5,6] or low order approximation attacks [7,8]. Such cryptographic weaknesses are in turn strongly related to the cryptographic properties of the Boolean functions that generate the corresponding sequences (see, e.g. Ref. [9] for an excellent reference on cryptographic Boolean functions). Therefore, a research question that naturally appears is the following: given two de Bruijn sequences y and $\tilde{y}$ of the same order n with a common subsequence s such that that there is no other de Bruijn sequence y' sharing with y a larger common subsequence than s, what is the relationship between the cryptographic properties (i.e. algebraic degree, nonlinearity, algebraic immunity, etc.) of the two corresponding Boolean functions that generate y and $\tilde{y}$, respectively? The same question can be generalized to cover cases in which a large (not necessarily the largest) subsequence is being shared between y and y'.

This chapter focuses on the above problem toward presenting some new results with the aim of paving the way for further research. More precisely, using as vehicle some recent results regarding the properties of the suffix arrays of de Bruijn sequences [10], we discuss how (inverse) suffix arrays may facilitate the finding of a cross-join pair in y with the (possibly) maximum diameter. By these means, we develop an approximation algorithm that applies directly to the inverse suffix array of y in order to provide an (almost) optimal solution to the above problem (whereas ensured optimality of the solution can be also proved in some cases). Next, based on this algorithm, some experiments are being performed to assess the properties of the corresponding Boolean functions that are associated with each such sequence. The results from this preliminary study illustrate that, even for small values of order n ($5 \leq n \leq 9$), there may exist some deviations in the corresponding cryptographic properties. This implies that, in the process of finding a cryptographic Boolean function f_y for generating a de Bruijn sequence y, it is also of importance to assess — toward evaluating the overall strength of f_y — the properties of the Boolean function $f_{\tilde{y}}$ which generates the de Bruijn sequence $\tilde{y}$, as described previously; this could be actually stated as a new cryptographic criterion for Boolean functions generating de Bruijn sequences.

This chapter is organized as follows. First, in Section 2, background information on cryptographic sequences and Boolean functions is given, identifying their relationships, whereas special emphasis on cryptographic properties of Boolean functions is given;

de Bruijn sequences and their properties are also explicitly discussed, whereas known results on the properties of functions generating de Bruijn sequences are also reviewed. Section 3 presents an approximation algorithm for computing, for a given de Bruijn sequence y, another one sharing a long (possibly the longest possible) subsequence with the initial. Therefore, having this algorithm as a tool to allow computing pairs of de Bruijn sequences with a long common subsequence, some experimental results on the relevant cryptographic properties of the underlying Boolean functions that are associated with such pairs of sequences are given in Section 4. Finally, concluding remarks and discussion on further research are given in Section 5.

2.   Background

2.1.   *Sequences and Boolean functions*

Binary sequences are widely used in cryptographic applications (see, e.g. Ref. [11], being a classical source for cryptography). A main representative of cryptographic systems that is strongly based on binary sequences is the so-called *stream ciphers*, which are usually preferred in highly restricted computing environments with the need for high speed encryption — and, thus, they gain much attention for, e.g. Internet of Things (IoT) applications. In stream ciphers, a pseudo-random keystream $k = k_1 k_2 \ldots$ is xor-ed with the original message (plaintext) $m = m_1 m_2 \ldots$ resulting in the encrypted message (ciphertext) $c = c_1 c_2 \ldots$ that satisfies $c_i = m_i \oplus k_i$ for all i (see Figure 1).

More generally, a binary sequence is a sequence $y = \{y_i\}_{i \geq 0}$ with elements in the finite field $\mathbb{F}_2$. We denote by y_i^j, with $i \leq j$, the

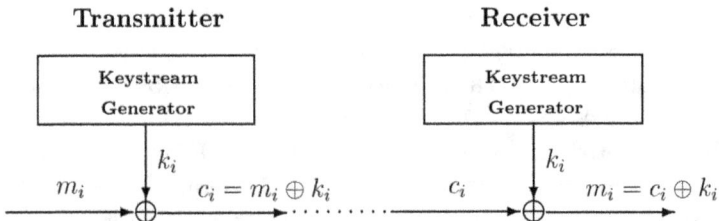

Transmitter	Receiver

Keystream Generator   Keystream Generator

$$m_i \quad \xrightarrow{\quad k_i \quad} \quad c_i = m_i \oplus k_i \quad \cdots \cdots \cdots \quad c_i \quad \xrightarrow{\quad k_i \quad} \quad m_i = c_i \oplus k_i$$

Figure 1.   The operation of a stream cipher.

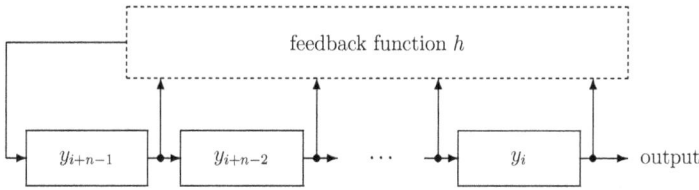

Figure 2. A feedback shift register.

subsequence (tuple) $y_i \ldots y_j$. The sequence y is said to be *ultimately periodic* if there exist integers $N > 0$ and $t \geq 0$ such that $y_{i+N} = y_i$ for all $i \geq t$. The least integer N with this property is called *period* of y, and t is its *preperiod*; if t equals zero, then the sequence y is said to be *periodic*. Such sequences are typically generated by feedback shift registers (FSRs), satisfying a recurring relation of the form

$$y_{i+n} = h(y_{i+n-1}, \ldots, y_i), \quad i \geq 0, \tag{1}$$

where $n \in \mathbb{Z}$ determines the number of stages of the register (Figure 2).

The feedback $h : \mathbb{F}_2^n \to \mathbb{F}_2$ is a function mapping elements of the nth-dimensional vector space $\mathbb{F}_2^n$ onto $\mathbb{F}_2$, i.e. it is a *Boolean* function with n variables. Let us denote by $\mathbb{B}_n$ the set of Boolean functions $f : \mathbb{F}_2^n \to \mathbb{F}_2$ on n variables as well as $[n] \triangleq \{1, \ldots, n\}$. Then, any $f \in \mathbb{B}_n$ can be uniquely expressed in its *algebraic normal form* (ANF) as follows [12]:

$$f(\vec{x}) = \sum_{I \subseteq [n]} \alpha_I \, \vec{x}^I, \quad \alpha_I \in \mathbb{F}_2, \tag{2}$$

where $\vec{x} = (x_1 \cdots x_n) \in \mathbb{F}_2^n$, $\vec{x}^I = \prod_{i \in I} x_i$ (by convention $\vec{x}^{\emptyset} = 1$) and the sum is taken modulo 2, i.e. over $\mathbb{F}_2$. The *algebraic degree* (or simply *degree*) of f, denoted by $\deg(f)$, is the highest number of variables that appear in a monomial of its ANF. If $\deg(f) = 1$, then f is a *linear* function. The binary vector of length 2^n

$$\vec{f} = (f(0 \cdots 0) \; f(0 \cdots 1) \; \cdots \; f(1 \cdots 1))$$

is the truth table of $f \in \mathbb{B}_n$ and the set $\operatorname{supp}(f) = \{\vec{x} \in \mathbb{F}_2^n : f(\vec{x}) \neq 0\}$ is called its *support* (defined for vectors in a similar manner). It is well-known that if $\deg(f) \leq r$, then $\vec{f}$ is a codeword of

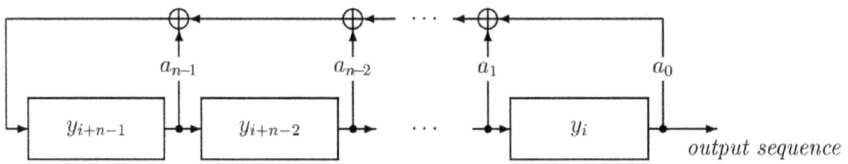

Figure 3.   An LFSR generating $y_{i+n} = \sum_{\ell=0}^{n-1} \alpha_\ell y_{i+\ell}$.

the rth order binary Reed–Muller code $\mathrm{RM}(r, n)$ [12]. The function $f \in \mathbb{B}_n$ is said to be *balanced* if its Hamming weight $\mathrm{wt}(f) = |\mathrm{supp}(f)|$ is equal to 2^{n-1}. Note that $\mathrm{wt}(f)$ is odd if and only if $\deg(f) = n$.

In general, the security of stream ciphers is strongly contingent on the pseudorandomness of the sequences that are employed as keystreams. The pseudorandomness of a sequence is attributed to several factors. Among others, an important cryptographic feature of a sequence is its *linear complexity*, defined as the length of the shortest Linear Feedback Shift Register (LFSR) — i.e. a register with a linear feedback function (see Figure 3, where each α_i, $i = 0, 1, \ldots, n-1$ indicates whether the corresponding linear term is present in the ANF of the feedback function) that can generate the sequence. The linear complexity of a sequence actually represents the minimum amount of the sequence required to fully specify the remainder; if a sequence has linear complexity L, then knowledge of $2L$ consecutive bit suffices to fully estimate the whole sequence through the prominent Berlekamp–Massey Algorithm (BMA) [13,14].

Therefore, a large linear complexity is prerequisite for keystreams. To achieve this, nonlinear Boolean functions are employed as building primitives for keystream generators. Typical classes of keystream generators are the cases of nonlinear filter (Figure 4(a)) and nonlinear combination generators (Figure 4(b)), which utilize nonlinear Boolean functions to one (the former case) or to many (the latter case) LFSRs. LFSRs are traditionally being chosen due to their nice mathematical properties, for example, it is known how to choose an LFSR producing sequences with the maximum possible period $2^n - 1$, i.e. that passes through all possible states, with the exemption of the all-zeroes state, while additionally LFSRs are fast and easy to be implemented (see, e.g. Ref. [15] for a classical source on this topic). Linear complexity has been studied to a great extent in the literature, employing, among others, construction of nonlinear systems that

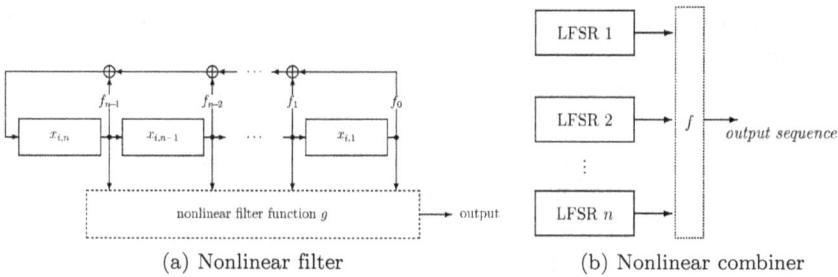

(a) Nonlinear filter   (b) Nonlinear combiner

Figure 4. Keystream generators based on nonlinear functions applied to LFSRs.

generate sequences with guaranteed high linear complexity, ensuring well-determined lower bounds (see, e.g. Refs. [16–20]).

During the recent years, the designs of keystream generators are focused on NLFSRs, which are being considered as more powerful in terms of thwarting specific types of attacks, such as algebraic attacks [21]. Notable examples of modern stream ciphers that utilize NLFSRs are the Grain cipher [2], being one candidate for standardization in the NIST competition for lightweight cryptography, as well as the Espresso cipher [22]. However, there are still several open questions in terms of designing NLFSRs with ensured properties, for example, it is still unknown how to construct, for the general case, a feedback function that allows an NLFSR to generate a sequence with the maximum possible period.

Regardless of how a nonlinear Boolean function is being used in a cryptographic system, there are several well-determined cryptographic criteria of Boolean functions that are desirable; more precisely, if a Boolean function does not possess one such criterion, then — depending on how the function is employed in the cryptosystem — a specific cryptanalytic attack may be efficiently mounted. For example, we already discussed that a high *algebraic degree* is essential to address BMA-based attacks, whereas *balancedness* is also clearly essential to ensure an unbiased output sequence. Other important cryptographic criteria are being discussed next.

Definition 1. We say that $f \in \mathbb{B}_n$ is *tth correlation immune* if it is not correlated with any t-subset of $\{x_1, \ldots, x_n\}$, namely, if

$$\Pr(f(\vec{x}) = 0 | x_{i_1} = b_{i_1}, \ldots, x_{i_t} = b_{i_t}) = \Pr(f(\vec{x}) = 0)$$

for any t positions $x_{i_1}, \ldots, x_{i_t}$ and any $b_{i_1}, \ldots, b_{i_t} \in \mathbb{F}_2$.

If a tth order correlation immune function is also balanced, then it is called *tth order resilient* [23].

The correlation immunity has been introduced to evaluate the resistance of a function against a known type of attacks that can be applied to stream ciphers, being called *correlation attacks*, first introduced in Ref. [23] for nonlinear combiners (but can be also applied, suitably modified, to nonlinear filter generators [24]). The correlation attack exploits the existence of a statistical dependence between the keystream and the output of a single constituent LFSR. Many variations of correlation attacks occur, with the *fast correlation attacks* being a notable case [25].

There is a known trade-off though between the correlation immunity and the degree of a function [23]: if $f \in \mathbb{B}_n$ is kth order correlation-immune, then $\deg(f) \leq n - k$. More specifically, for a resilient function f of order k, it holds $\deg(f) \leq n - k - 1$ if $1 \leq k \leq n - 2$ [23].

Definition 2. The nonlinearity of $f \in \mathbb{B}_n$ is defined to be

$$\mathrm{nl}(f) = \min_{g \in \mathrm{RM}(1,n)} \mathrm{wt}(f + g).$$

The nonlinearity, as a cryptographic criterion, stems from the fact that if a high-degree cryptographic Boolean function f can be adequately approximated by a linear function g, then the corresponding cryptographic system is vulnerable to specific attacks exploiting this resemblance between f and g, for instance, the linear cryptanalysis on stream ciphers [26] exploits the existence of biased linear relations between some keystream bits and some key bits.

Apart from the linear approximations, cryptographic Boolean functions should not be well approximated by other low-degree function (especially from functions with degree $r \leq 2$). Such approximations could allow other more sophisticated, attacks; for example, the attacker may replace, in a nonlinear combiner generator, the corresponding Boolean function f by another function g of low degree such that $\mathrm{wt}(f + g)$ is small [8] and, thus, able to mount a so-called fast correlation attack to the modified system in order to reveal a secret key which produces a slightly modified keystream (namely, a noisy version of the keystream, where the noise is small). Hence, the notion of the nonlinearity is readily generalized to the *rth order* *nonlinearity* $\mathrm{nl}_r(f)$ of the function f [7], which is defined as

$$\mathrm{nl}_r(f) = \min_{g \in \mathrm{RM}(r,n)} \mathrm{wt}(f+g). \tag{3}$$

Computing the rth order nonlinearity of Boolean functions, as well as their best rth order approximations, is known to be a difficult task even for small values of r. For the linear case, i.e. for $r = 1$, the maximum possible nonlinearity is achieved by the so-called *bent* functions, which is $2^{n-1} - 2^{n/2-1}$ for n even; if n is odd, then the value of the maximum possible achievable nonlinearity remains unknown. However, bent functions are not balanced and the maximum possible nonlinearity of a balanced Boolean function is also still unknown for the general case (only for small values of n, the maximum attainable nonlinearity is known based on exhaustive computations). For $r \geq 2$, some results for specific cases are only known (see, e.g. Ref. [27–31]).

Very recently [32], it has been proved that, by defining a bijection between 2^n-periodic binary sequences and Boolean functions on n variables, a generalization of the linear complexity being called *error linear complexity spectrum* — which indicates how the linear complexity of a sequence can be decreased if some of its elements are modified — also provides useful cryptographic information for the corresponding Boolean function f, namely, it yields an upper bound on the minimum Hamming distance between f and the set of functions depending on fewer number of variables. As it is also shown in Ref. [32], low-degree approximations of a Boolean function can also be obtained in an efficient way through this approach.

Definition 3. A function is called an annihilator of $f \in \mathbb{B}_n$ if and only if it belongs to

$$A_f = \{g \in \mathbb{B}_n : fg = 0\}.$$

Definition 4. The algebraic immunity $\mathrm{AI}_n(f)$ of $f \in \mathbb{B}_n$ is the least degree of all the nonzero annihilators of f and $1 + f$.

A cryptographic Boolean function f should have high AI to be resistant against a specific type of attacks, being called algebraic attacks [21,33]. A well-known result is that $\mathrm{AI}_n(f) \leq \lceil \frac{n}{2} \rceil$ for all $f \in \mathbb{B}_n$ [21]. It is also well known that low nonlinearity implies low algebraic immunity (although high nonlinearity does not always imply high algebraic immunity). More precisely, the following holds [34]:

$$\text{nl}_r(f) \geq \sum_{i=0}^{\text{AI}_n(f)-r-1} \binom{n}{i}.$$

The lower bound of the rth order nonlinearity of algebraic immune functions has been improved, for several cases, in Refs. [35–37].

A powerful version of algebraic attacks is the so-called fast algebraic attacks [38]; a maximum algebraic immunity, although it is prerequisite, does not suffice to ensure resistance against fast algebraic attacks. For this case, the relative cryptographic criterion is the so-called *fast algebraic immunity*, which is traditionally defined as follows [39].

Definition 5. The fast algebraic immunity of $f \in \mathbb{B}_n$, denoted by $\text{FAI}_n(f)$, is defined as

$$\text{FAI}_n(f) = \min_{g \in \mathbb{B}_n}\{2\,\text{AI}_n(f), \deg(g) + \deg(fg)\},$$

where $1 \leq \deg(g) \leq \text{AI}_n(f)$.

It can be easily verified that $\text{FAI}_n(f) \leq n$, while any function with maximum FAI has also maximum AI, but the converse does not hold. More recently, a slightly modified criterion of FAI has been introduced in Ref. [40], whereas it is also shown that both the algebraic immunity as well as the fast algebraic immunity can be expressed by means of the so-called *immunity profile*.

There are many constructions in the literature of functions achieving maximum algebraic immunity; however, not all of them behave well with respect to fast algebraic attacks. Even for those functions that seem to have good behavior for fast algebraic attacks, this has been mainly shown through experiments for relative small values of n. A classical such class of functions, in which the maximum algebraic degree also has high nonlinearity, is presented in Ref. [41], whereas several other constructions based on modifications of this class are also known (see, e.g. Refs. [42–45]). The class of functions described in Ref. [41] has been (later on) mathematically proved to have maximum fast algebraic immunity, i.e. any such function is a *perfect algebraic immune* function in Ref. [46]. The fast algebraic immunity is being currently studied by many researchers, being a highly active research field (see, e.g. Ref. [47–49]).

Note that, due to the aforementioned trade-off between correlation immunity and algebraic degree, there is also a trade-off between fast algebraic immunity and correlation immunity; indeed, if $f \in \mathbb{B}_n$ is mth order correlation-immune Boolean function for $m > 2$, then it does not have maximum fast algebraic immunity [50].

2.2. *de Bruijn sequences*

A binary de Bruijn sequence of order n is a 2^n-periodic sequence that contains each n-tuple exactly once within one period. For example, for $n = 4$, such a de Bruijn sequence of period $2^4 - 1 = 15$ is the following:

$$y = 0000100110101111\ldots.$$

By considering, as sequence generators, any possible finite-state machine, we get that de Bruijn generators are fully characterized by well-determined properties in terms of controllability and observability conditions of the system [51].

Clearly, any NLFSR of length n which passes through all possible 2^n states — including the all-zeroes state — generates a de Bruijn sequence. Apparently, if h is the feedback function of an NLFSR that generates de Bruijn sequence, it holds $h(0, 0, \ldots, 0) = 1$ (to avoid the all-zeroes cycle) and $h(1, 1, \ldots, 1) = 0$ (to avoid the all-ones cycle). This means that the ANF of h has a nonzero (i.e. the value 1) as constant term, whereas the number of all the monomials in the ANF of h, including the constant term, is even. However, appropriately choosing a feedback function h for NLFSR such that it generates a de Bruijn sequence is still an open problem, since only a few such feedback functions are known [3].

There are many combinatorial algorithms for constructing de Bruijn sequences. A classical method is proposed by Lempel in Ref. [52] which, having a de Bruijn sequence of order n, a de Bruijn sequence of order $n+1$ is obtained through a specific procedure based on a mapping from $\mathbb{F}_2^n$ to $\mathbb{F}_2^n$ called D-morphism. In Ref. [53], it is shown how the construction of Ref. [52] can be used so as to produce a maximum-period NLFSR from a shorter-stage maximum-period

NLFSR; several other constructions are also based on the same concept (see, e.g. Refs. [54,55]). Other approaches also exist, see, e.g. Ref. [56].

A popular approach for constructing new de Bruijn sequences from given known Dr Bruijn sequences is based on splitting a cycle (i.e. a sequence of 2^n states of an NLFSR generating a de Bruijn sequence) into two cycles and subsequently appropriately joining them into forming new cycle (i.e. a new sequence of states, which yields another de Bruijn sequence). Two distinct such pairs are employed for these two cases (splitting and joining) that are called *cross-join pairs*. To define the notion of cross-join pairs, we follow the notation of Ref. [57]; let $y = y_0 y_1 \ldots y_{2^n-1}$ be a de Bruijn sequence. For any n-tuple $\vec{a} = (a_0 a_1 \ldots a_{n-1}) \in \mathbb{F}_2^n$, we define its conjugate $\hat{\vec{a}} = (a_0' a_1 \ldots a_{n-1}) \in \mathbb{F}_2^n$, where $a_0' = a_0 + 1$. Clearly, any $\vec{a} \in \mathbb{F}^n$ is uniquely associated with a state of maximum-period NLFSR of length n.

Definition 6. Two conjugate pairs $(\vec{a}, \hat{\vec{a}})$ and $(\vec{b}, \hat{\vec{b}})$ constitute a cross-join pair for the sequence y if these four n-tuples occur, under a cyclic shift of y, in the order

$$\ldots \vec{a}, \ldots, \vec{b}, \ldots, \hat{\vec{a}}, \ldots, \hat{\vec{b}} \ldots$$

within y, i.e. $\vec{b}$ appears after $\vec{a}$, $\hat{\vec{a}}$ occurs after $\vec{b}$ and $\hat{\vec{b}}$ occurs after $\hat{\vec{a}}$.

Hence, interchanging the successors of $\vec{a}, \hat{\vec{a}}$ results in splitting y into two cycles, i.e. into new distinct sequences of smaller lengths. Next, we join these two cycles into a new one by interchanging the successors of $\vec{b}, \hat{\vec{b}}$. The significance of using cross-join pairs rests with the following result.

Theorem 1 ([57]). *Let y, w be two de Bruijn sequences of order n. Then w can be obtained from y by repeated application of the cross-join pair operation.*

It is still unknown which is the number of cross-join pairs in an arbitrary de Bruijn sequence. For the special case of a sequence generated by an LFSR passing through all the possible states with the exception of the all-zeroes state (i.e. a maximal-length sequence, which is a specific type of the so-called *modified de Bruin sequences*), it has been proved that the number of cross-join pairs in an n-bit maximum-length LFSR is $(2^{n-1} - 1)(2^{n-1} - 2)/6$ [58].

There are several other algorithmic approaches to construct de Bruijn sequences from the scratch. The following is one of the easiest [59]: start with n zeros and append a 1 whenever the n-tuple thus formed has not appeared in the sequence so far, otherwise append a 0. This technique is widely known as the *prefer-one algorithm*. Other implementations of de Bruijn generators are given in Refs. [60–63].

It is well known that the linear complexity of a de Bruijn sequence of order n takes values between $2^{n-1} + n$ and $2^n - 1$. Both the upper and lower bounds are achievable, as it is proved in Refs. [64,65], respectively. Some results on the linear complexity of any de Bruijn sequence are summarized in Ref. [66].

Several open questions still exist with regard to the cryptographic properties of Boolean functions that generate de Bruijn sequences. The nonlinearity criterion is studied in Ref. [67]. The following result is proved therein.

Proposition 1. *Let* $f \in \mathbb{B}_n$ *a Boolean function generating a de Bruijn sequence* y *of order* $n > 2$. *Then, it holds* $\mathrm{nl}(f) \leq 2^{n-1} - 2^{\frac{n-1}{2}}$.

In Ref. [67], some specific constructions of de Bruijn generators are given, which achieve nonlinearity equal to 2, i.e. the least possible. Moreover, in Ref. [67], some techniques are described to increase the nonlinearity of a given feedback function using cross-joining.

2.2.1. *Suffix arrays of de Bruijn sequences*

More recently, the properties of suffix arrays of de Bruijn sequences have been studied in Ref. [10]. Suffix arrays constitute important indexing data structure for sequences and are defined as follows.

Definition 7 ([68]). The *suffix array* S of a binary sequence $y^N \triangleq y_0 y_1 y_2 \cdots y_{N-1}$ holds the starting positions (ranging from 0 to $N-1$) of its N lexicographically ordered suffixes, i.e. for $0 \leq i < j < N$, it holds $y_{S[i]}^{N-1} < y_{S[j]}^{N-1}$.

Apart from the suffix array, the *inverse suffix array* S^{-1} is also frequently used, namely, $S^{-1}[i]$ indicates the number of suffixes that are lexicographically smaller than y_i^{N-1}; by definition, $S[S^{-1}[i]] = i$ for all $i = 0, 1, \ldots, N - 1$. Both data structures have size $\mathcal{O}(N \log N)$ and can be constructed in linear time (see, e.g. Ref. [69]).

In Ref. [10], by slightly differentiating the definition of suffix arrays so as to allow them to describe any de Bruijn sequence y which is periodic and not finite-length (namely, the sequence y_0^{N+n-1}, i.e. the whole period of y augmented with the first n bits of its period, based on the repetition that occurs, is being considered), specific properties of the suffix array of a de Bruijn sequence are proved. Due to these properties, it is shown that the construction of a de Bruijn sequence is somehow equivalent with a construction of a suffix array satisfying these properties, since any array with these properties is uniquely associated with a specific de Bruijn sequence. Moreover, it is shown therein that inverse suffix arrays facilitate, as indexing structures, the identification of cross-join pairs within a de Bruijn sequence. More precisely, the following holds.

Lemma 1 ([10]). *Let y be a binary de Bruijn sequence of order n and let*

$$(y_{m_1}^{m_1+n-1}, y_{m_2}^{m_2+n-1}) \quad and \quad (y_{\ell_1}^{\ell_1+n-1}, y_{\ell_2}^{\ell_2+n-1})$$

be a cross-join pair, where $y_{m_2}^{m_2+n-1}$, $y_{\ell_2}^{\ell_2+n-1}$ are the conjugates of $y_{m_1}^{m_1+n-1}$, $y_{\ell_1}^{\ell_1+n-1}$, respectively. Then, m_1, m_2, ℓ_1, ℓ_2 occur in the following order:

$$\ldots\ m_1\ \ldots\ \ell_1\ \ldots\ m_2\ \ldots\ \ell_2\ \ldots$$

within S^{-1}, where S^{-1} is the inverse suffix array of y.

Moreover, having the inverse suffix array of a de Bruijn sequence y renders the identification of conjugate pairs trivial due to the following.

Proposition 2 ([10]). *Let y be a de Bruijn sequence of order n. Then, the tuple y_m^{m+n-1} is the conjugate of $y_{m'}^{m'+n-1}$, for $m \neq m'$, if and only if it holds*

$$|S^{-1}[m] - S^{-1}[m + 1]| = 2^{n-1}.$$

Example 1 ([10]). Let us consider the de Bruijn sequence $y = 0000101001101111\ldots$ of order $n = 4$ ($N = 2^n = 16$). It can be easily verified that its suffix array S is equal to

$$S = (0\ 1\ 2\ 7\ 5\ 3\ 8\ 11\ 15\ 6\ 4\ 10\ 14\ 9\ 13\ 12).$$

Indeed, $y_0^{19} < y_1^{19} < y_2^{19} < y_7^{19} < y_5^{19} < \cdots < y_{13}^{19} < y_{12}^{19}$ (where the ordering is lexicographical).

In addition, it is also easily verified that its inverse suffix array S^{-1} is equal to

$$(0\ 1\ 2\ 5\ 10\ 4\ 9\ 3\ 6\ 13\ 11\ 7\ 15\ 14\ 12\ 8).$$

According to Proposition 2, the conjugate pairs are indexed as follows:

$$(0,8), (1,9), (2,10), (3,11), (4,12), (5,13), (6,14), (7,15).$$

Examining for cross-join pairs within S^{-1} according to Lemma 1, we see that such a cross-join pair is, e.g. $(1,9), (5,13)$. Hence, interchanging the successors of these pairs (i.e. the successors of 1, 9 and the successors of 5, 13), we get a new array

$$(0\ 1\ 3\ 6\ 13\ 10\ 4\ 9\ 2\ 5\ 11\ 7\ 15\ 14\ 12\ 8).$$

It is easy to see that the latter is the inverse suffix array of the de Bruijn sequence $y' = 0000110100101111\ldots$.

2.2.2. *de Bruijn sequences sharing the longest common subsequence*

As also stated in Section 1, the problem of finding, for a given de Bruijn sequence y of order n, another de Bruijn sequence $\tilde{y}$ with the same order n having the longest possible common subsequence with y has been recently investigated in Ref. [4]. The main results therein rest with the fact that such a sequence $\tilde{y}$ can be obtained through an appropriate single cross-join operation from y. To this end, we first present the notion of the *diameter* of a sequence, as it is given in Ref. [4].

Definition 8 ([4]). For a given cross-join pair of a de Bruijn sequence y cutting the cycle to four subsequences, the maximum length of the four subsequences is defined to be the diameter of the cross-join pair.

The aforementioned notion of the diameter of a cross-join pair is essential in determining the sequence $\tilde{y}$ with the longest common subsequence with y, as proved in Ref. [4].

Theorem 2 ([4]). *Let y be a de Bruijn sequence of order n. Assume $\tilde{y}$ is a de Bruijn sequence sharing the longest possible subsequence with y. Then $\tilde{y}$ is generated from y by a single cross-join operation.*

Due to Theorem 2, the problem of determining the longest sharing subsequence is reduced to finding cross-join pairs with the biggest diameter.

Some bounds on the lengths of the longest common subsequences between two de Bruijn sequences are also proved in Ref. [4]. More precisely, for any de Bruijn sequence y of order $n \geq 3$, there exists a de Bruijn sequence of order n sharing a subsequence of length at least $2^{n-1} + n - 3$ with it. Moreover, for $n \geq 3$, there exists a de Bruijn sequence sharing a subsequence of length $2^n - 3$ with the so-called *prefer-one sequence*. Finally, for $n \geq 5$, there exist two de Bruijn sequences with order n sharing a subsequence of length $2^n - 2$ (while this cannot hold for $n = 3$ or $n = 4$); this result is based on a construction method for deriving, for any such n, a de Bruijn sequence containing a cross-join pair with diameter $2^n - n - 1$ [4].

3. de Bruijn Sequences Sharing the Longest Common Subsequence Revisited

In this section, we shall investigate the method of determining a cross-join pair of a de Bruijn sequence, having the largest diameter. To this end, specific properties of the inverse suffix array of a de Bruijn sequence y are being used. First, we can easily prove the following:

Proposition 3. *Let y be a binary de Bruijn sequence of order n and S^{-1} its inverse suffix array. Then, of any $0 \leq i \leq 2^n - 1$, it holds $S^{-1}[i] = k$, where k is the integer whose binary representation coincides with the n-tuple y_i^{i+n-1}.*

Proof. By the definition, $S^{-1}[i]$ indicates the number of suffixes that are lexicographically smaller than y_i^{N-1}. Since each n-tuple appears exactly once within the period of a de Bruijn sequence, it is clearly equivalent to state that, for a de Bruijn sequence, $S^{-1}[i]$ indicates the number of suffixes that are lexicographically smaller than y_i^{i+n-1}. By associating each n-tuple y_i^{i+n-1}, $i = 0, 1, \ldots, 2^n - 1$, with the corresponding decimal number

$$I = 2^0 * y_{i+n-1} + 2^1 * y_{i+n-2} + \cdots + 2^{n-1} * y_i,$$

i.e. I is the integer whose binary representation is given by the n-tuple y_i^{i+n-1}, we get that all the possible 2^n n-tuples appearing in a de Bruijn sequence cover the entire space of integers $\{0, 1, \ldots, 2^n - 1\}$ with cardinality 2^n and, moreover, the lexicographical ordering of n-tuples is equivalent to the classical ordering of the corresponding decimal representations. Hence, we conclude that, for any $0 \leq i \leq 2^n - 1$, the number of suffixes that are lexicographically smaller than y_i^{N-1} is equal to I, and thus, the claim follows. $\square$

Example 2. Let us recall the de Bruijn sequence $s = 0000101001101111$ from Example 1. As it has been shown, its inverse suffix array is

$$S^{-1} = (0 \ 1 \ 2 \ 5 \ 10 \ 4 \ 9 \ 3 \ 6 \ 13 \ 11 \ 7 \ 15 \ 14 \ 12 \ 8).$$

By observing y and S^{-1}, the property implied by Proposition 3 is easily seen. For example, $S^{-1}[3] = 5$, whose binary representation is 0101, which in turn coincides with y_3^6.

A direct outcome from Proposition 3 is the following.

Corollary 1. *Let y, $\tilde{y}$ be two distinct de Bruijn sequences of order n, with S^{-1} and $\tilde{S}^{-1}$ being their corresponding inverse suffix arrays. Then, if S^{-1} and $\tilde{S}^{-1}$ coincide at r consecutive places (possibly up to a proper cyclic shift of the sequences), i.e. there exists $0 \leq i \leq 2^n - 1$ such that*

$$S^{-1}[(i + m) \pmod{2^n}] = \tilde{S}^{-1}[(i + m) \pmod{2^n}], \quad \forall 0 \leq m < r,$$

then y and $\tilde{y}$ share a common subsequence with length $r + n - 1$.

Proof. From Proposition 3, our hypothesis implies that

$$y_{i+m}^{i+m+n-1} = \tilde{y}_{i+m}^{i+m+n-1}, \quad \forall 0 \leq m < r,$$

which is equivalent to state that $y_i^{i+r+n-2} = \tilde{y}_i^{i+r+n-2}$; thus, the claim follows. $\square$

Looking for conjugate pairs in a de Bruijn sequence s becomes more convenient when having S^{-1} due to Proposition 2. This in turn facilitates the task of finding the cross-pair with the largest diameter. In this direction, we can prove the following.

Theorem 3. *Let i, j be the indices of a de Bruijn sequence y of order n corresponding to a conjugate pair, whereas k, ℓ the indices of y of another conjugate pair so that these two pairs constitute a cross-join pair. Without loss of generality, we assume that y is in a cyclic shift such as $\ell = 2^n - 1$. Then, this cross-join pair has the largest diameter if and only if the distance $\ell - i$ is the least possible among all cross-join pairs of y — and this holds for any possible cyclic shift of y.*

Proof.   The cross-join pair in our hypothesis results in splitting y into four subsequences, which appropriately recombined yield another de Bruijn sequence $\tilde{y}$. By considering the corresponding inverse suffix arrays S^{-1} and $\tilde{S}^{-1}$ of y and $\tilde{y}$, respectively, it is readily seen that $S^{-1}[p] = \tilde{S}^{-1}[p]$ for all $0 \le p \le i$; indeed, for this part of S^{-1} no interchanging of values takes place toward constructing $\tilde{S}^{-1}$ through the cross-join pair operation (the first such interchange occurs just after $S^{-1}[i]$). Hence, recalling Corollary 1, y and $\tilde{y}$ coincide at their first $i + n$ places, i.e. $y_0^{i+n-1} = \tilde{y}_0^{i+n-1}$.

Let us assume that there is another cross-join pair, determined by the indices i', j' (for the first conjugate pair) and k', ℓ' (for the second conjugate pair) having larger diameter than the one induced by i, j and k, ℓ. Without loss of generality, we consider the appropriate cyclic shift $\hat{y}$ of y such as its largest subsequence of y obtained through the application of this cross-join pair operation on y is a prefix of $\hat{y}$. Let r be the length of this subsequence, i.e. the diameter of y. Then this shifted version $\hat{y}$ of y can be written as

$$\hat{y} = \hat{y}_0 \hat{y}_1 \cdots \hat{y}_{r-1} \hat{y}_r \hat{y}_{r+1} \cdots \hat{y}_{2^n-1}. \tag{4}$$

Since r is a diameter, it clearly holds $r > i + n$ since, otherwise, the splitting induced by the cross-join pair implied by i, j and k, ℓ would yield a larger subsequence. Moreover, the initial indices i', j', k', ℓ' correspond, due to the shifting operation, in the part $\hat{y}_r^{2^n-1}$, whose length is smaller than the length of $y_{i+n}^{2^n-1}$, which contradicts our hypothesis that the distance $\ell - i$ is the least possible among all cross-join pairs of y. Thus, the claim follows.   □

Based on Theorem 3, we present an approximation algorithm (see Algorithm 1) that aims to find a cross-join pair with a large (possibly the largest) diameter; the main idea is to first find out a conjugate pair (indexed by i, j) with the smallest possible distance w between the corresponding indices (see line 2 of the algorithm) and, next, to find out another conjugate pair (indexed by k, ℓ) so as to constitute (with the initial one) a cross-join pair and, moreover, this new pair satisfies, for the given initial pair i, j, the following property: the distance between i and ℓ is the least possible (see lines 10–15 of the algorithm). Note that, for the original conjugate pair i, j, the actual distance w between i and j should be considered under any cyclic shift of y — that is why we compute not only the difference $j - i$ but also the difference $2^n - (j - i)$ (and, clearly, w is the minimum of these two values). Once we find out such a pair, it is convenient to consider the shifted version of the sequence starting by its ith

---

Algorithm 1. Find the cross-join pair with the (approximately) largest diameter in a de Bruijn sequence

---

Require: inverse suffix array S^{-1} of a de Bruijn sequence y with order n

1: $d \leftarrow 2^n$

2: $\{i, j\} \triangleq \min_{0 \le i < j < 2^n - 1} \min\{(j - i), 2^n - (j - i)\} : |S^{-1}[i] - S^{-1}[j]| = 2^{n-1}, j > i + 1$

3: **if** $2^n - (j - i) < j - i$ **then**

4: $\quad w = 2^n - (j - i)$

5: $\quad t \leftarrow i; i \leftarrow j; j \leftarrow t$ ▷ Swap the values of i, j

6: **else**   ▷ i.e. $j - i \le 2^n - (j - i)$

7: $\quad w = j - i$

8: **end if**

9: $S^{-1} \leftarrow$ shifted version of S^{-1} by i positions to the left

10: **for** $k \leftarrow 1 : w - 1$ **do** ▷ The conjugate pair has indices $(0, w)$ in this shifted version

11: $\quad$ Find $\ell' > w : |S^{-1}[\ell'] - S^{-1}[k]| = 2^{n-1}$

12: $\quad$ **if** $\ell' - i < d$, **then**

13: $\quad\quad \ell \leftarrow \ell'$

14: $\quad\quad d \leftarrow \ell - i$

15: $\quad$ **end if**

16: **end for**

Ensure: i, w, k, ℓ such that, in the shifted version of y by i positions to the left, the conjugate pairs $(0, w)$ and (k, ℓ) constitute a cross-join pair with a large diameter

position, where i corresponds to the minimum value among the pair i, j and j corresponds to the maximum value such as $j - i = w$ (see line 9 of the algorithm in conjunction with the lines $3 - 5$).

Apparently, the above is an approximation algorithm since the derived solution is not bound to be always the optimal (namely, the algorithm does not perform an exhaustive search over all the possible cross-join pairs). However, for specific cases, we may be sure that the derived solution is the optimal, while in any case, it suffices to provide a "good" solution efficiently.

Corollary 2. *If Algorithm 1 returns i, w, k, ℓ such that $\ell = w + 1$, then this solution is the optimal (and not an approximate) solution.*

Proof. Straightforward from Theorem 3, taking into account that, according to Algorithm 1, w is the minimum distance between two indices corresponding to a conjugate pair. □

4. Exploring the Cryptographic Properties of the Relevant de Bruijn Generators

In this section, we present some experimental results on the cryptographic properties of Boolean functions generating de Bruijn sequences sharing a large (the largest possible in some cases) common subsequence. There results are shown in Table 1 (for $n = 6$ and $n = 7$) and in Table 2 (for $n = 8$ and $n = 9$). For each n, we first generated randomly, through the probabilistic procedure described in Ref. [10], 10 de Bruijn sequences (each of them is being called *initial* in the tables) and, then, by applying Algorithm 1, a new de Bruijn sequence that highly resembles the initial one is obtained; for each such pair (initial–new), the length of the longest common subsequence is given within parentheses in the new one. The tables present, for any such pair of de Bruijn sequences, the nonlinearity as well as the algebraic immunity of the corresponding de Bruijn generators; in cases that, for any such pair of sequences, either the nonlinearities or the algebraic immunities do not coincide, the lowest such value is written in bold to indicate this discrepancy.

Moreover, toward illustrating whether the relevant values should be considered as high or not, the maximum possible values for the

Table 1. Comparing de Bruijn generators with $n = 6$ and $n = 7$ variables.

de Bruijn ($n = 6$)	Nonlinearity (Max = 26)	AI (Max = 3)	de Bruijn ($n = 7$)	Nonlinearity (Max = 56)	AI (Max = 4)
Initial	18	3	Initial	**42**	3
New (60)	18	3	New (124)	46	3
Initial	22	3	Initial	46	3
New (60)	**18**	3	New (122)	46	3
Initial	22	3	Initial	42	3
New (58)	**18**	3	New (126)	42	3
Initial	**18**	3	Initial	46	3
New (58)	22	3	New (124)	**42**	3
Initial	14	3	Initial	46	3
New (60)	14	3	New (120)	46	3
Initial	22	3	Initial	46	3
New (58)	22	3	New (126)	46	3
Initial	**18**	3	Initial	46	3
New (60)	22	3	New (126)	46	3
Initial	22	3	Initial	**46**	3
New (60)	**18**	3	New (124)	50	3
Initial	**14**	3	Initial	46	3
New (58)	18	3	New (122)	**42**	3
Initial	**14**	3	Initial	42	3
New (60)	18	3	New (124)	**38**	3

nonlinearities (based on the best known nonlinearity for balanced Boolean functions with $n = 6, 7, 8, 9$ variables, according to Ref. [9]) and the algebraic immunities (i.e. $\lceil \frac{n}{2} \rceil$) are also shown in the headers of the columns. Note that these known upper bounds on the nonlinearities are tighter, for $n = 6, 7, 8, 9$, than the bound given in Proposition 1. For the computations of the cryptographic properties of the corresponding Boolean function, the free open-source SageMath tool has been used.

It should be stressed that we also computed the algebraic degrees of the functions; in all cases, they were found to be the maximum possible, i.e. $n-1$, and that is why they are omitted from the tables to simplify the presentation of the results. In addition, we also computed the linear complexities of all the sequences to see whether significant discrepancies occur between any two similar de Bruijn sequences; the relevant results are subsequently discussed.

Table 2. Comparing de Bruijn generators with $n = 8$ and $n = 9$ variables.

de Bruijn ($n = 8$)	Nonlinearity (Max = 116)	AI (Max = 4)	de Bruijn ($n = 9$)	Nonlinearity (Max = 240)	AI (Max = 5)
Initial	94	4	Initial	206	4
New (250)	**90**	4	New (508)	**202**	4
Initial	98	4	Initial	214	4
New (242)	**96**	4	New (504)	214	4
Initial	**98**	4	Initial	210	**4**
New (252)	102	4	New (510)	210	5
Initial	102	4	Initial	214	5
New (250)	**98**	4	New (504)	214	5
Initial	102	4	Initial	210	4
New (252)	**98**	4	New (506)	210	4
Initial	98	4	Initial	206	4
New (248)	98	4	New (510)	206	4
Initial	**98**	4	Initial	**206**	5
New (252)	102	4	New (506)	210	5
Initial	94	4	Initial	210	5
New (252)	94	4	New (506)	210	5
Initial	102	4	Initial	206	4
New (248)	**98**	4	New (506)	**204**	4
Initial	98	4	Initial	206	4
New (252)	98	4	New (508)	206	4

The main conclusions from the experimental results are the following:

(1) In about half the cases, for any n, there exist discrepancies in the nonlinearities between the generators producing similar de Bruijn sequences. These discrepancies are, in almost all the cases, of the maximum possible value, i.e. the nonlinearities differ by 4 (note that a cross-join pair operation results in modifying four entries in the truth table of the corresponding Boolean function). Moreover, the nonlinearities of the specific de Bruijn generators are generally not so close to the maximum possible achievable value by a balanced Boolean function; hence, it becomes evident that any further degradation in such cases — as those observed — may have an impact on the cryptographic strength of the function.

(2) The algebraic immunity seems to remain constant for any pair of generators producing similar de Bruijn sequences; for only one (out of 40) case, a decrease by 1 is observed. However, since it is well known that a difference by 1 in the algebraic immunity of a function, used as combiner or filter in a stream cipher, makes a big difference in the efficiency of algebraic attacks (see Ref. [9, p. 93]), even this single case indicates that algebraic immunity should also be considered when comparing functions generating similar de Bruijn generators. Moreover, interestingly enough, for $n = 7$ and $n = 9$, the vast majority of the de Bruijn generators that were randomly constructed achieve almost optimum (i.e. $\frac{n+1}{2} - 1$), but not optimum (i.e. $\frac{n+1}{2}$), algebraic immunity.

(3) The linear complexities between similar de Bruijn sequences are either identical or very close (and that is why we omitted them from the above tables); in cases where discrepancies occur in the linear complexities, these discrepancies take values from 1 (three cases for $n = 6$, seven cases for $n = 7$, four cases for $n = 8$ and three cases for $n = 9$) up to 4 (only one case for $n = 7$).

Although the above should be considered as a preliminary study, it becomes evident that de Bruijn generators providing similar de Bruijn sequences may have differences in their cryptographic criteria; such differences seem to be most probably on the corresponding nonlinearities, but also the algebraic immunities should be examined. Therefore, even from this preliminary study, we get that when assessing a de Bruijn generator in terms of its cryptographic properties, it is of interest to consider the properties of de Bruijn generators producing similar de Bruijn sequences. For example, for large values of n (e.g. for $n = 128$), a full-cycle NLFSR in a stream cipher is not expected to pass through all its states during an encryption process and, thus, it produces only a part of the corresponding de Bruijn sequence, which in turn may fully coincide with the corresponding part of a similar de Bruijn sequence obtained by a, let's say, "weaker" Boolean function; therefore, having any of these Boolean functions produces the same output in real cryptographic applications.

5. Conclusion

This chapter surveys some recent results on de Bruijn sequences, in conjunction with the cryptographic properties of de Bruijn generators, providing also some new results on establishing methods to find out de Bruijn sequences sharing a large (possibly the largest) common subsequence. To this end, inverse suffix arrays of these sequences are being used, which possess some nice properties. Furthermore, we present through some experiments that de Bruijn generators providing similar de Bruijn sequences may have important differences in their cryptographic properties, especially in the nonlinearity. Although these experiments are not extensive, the conclusions derived so far clearly illustrate that it is of cryptographic importance to consider, when assessing the properties of a de Bruijn generator, other de Bruijn generators that generate similar sequences.

This work opens several research directions; first, it is still an open question to find an efficient way to compute, for a given de Bruijn sequence, another sequence that is ensured to have the largest possible subsequence with the initial one (or, equivalently, to find cross-join pairs with the maximum diameter). Moreover, it is of high importance to establish (at least some) properties of the behavior of the cryptographic properties of a de Bruijn generator that is derived through a cross-join operation applied to another de Bruijn generator with known properties; this is of high relevance with the problem studied in this work. More generally, this work further reveals the need to have a better understanding of the cryptographic properties of de Bruijn generators. Some of them, such as the fast algebraic immunity, have not been studied at all so far in the literature.

References

[1] H. Fredricksen (1982). A survey of full length nonlinear shift register cycle algorithms. *SIAM Rev.*, **24**(2), 195–221.

[2] M. Hell, T. Johansson, A. Maximov, W. Meier, J. Sönnerup, and H. Yoshida (2021). Grain-128AEAD — A lightweight AEAD stream cipher. NIST. https://grain-128aead.github.io/.

[3] E. Dubrova (2012). A list of maximum period NLFSRs. *Cryptology ePrint Archive*, Report 2012/166. https://ia.cr/2012/166.

[4] Y. Jiang and D. Lin (2020). Longest subsequences shared by two de Bruijn sequences. *Des. Codes Cryptogr.*, **88**(7), 1463–1475.

[5] M. Matsui (1993). Linear cryptanalysis method for DES cipher. In *Advances in Cryptology — EUROCRYPT 93: Workshop on the Theory and Application of Cryptographic Techniques, Lofthus, Norway, May 23–27, 1993, Proceedings.* T. Helleseth (ed.) Lecture Notes in Computer Science, Vol. 765 (Springer), pp. 386–397.

[6] M. Hell and T. Johansson (2006). On the problem of finding linear approximations and cryptanalysis of Pomaranch version 2. In *Selected Areas in Cryptography: 13th International Workshop, SAC 2006, Montreal, Canada, August 17–18, 2006 Revised Selected Papers.* E. Biham and A. M. Youssef (eds.), Lecture Notes in Computer Science, Vol. 4356 (Springer), pp. 220–233.

[7] L. R. Knudsen and M. J. B. Robshaw (1996). Non-linear approximations in linear cryptanalysis. In *Advances in Cryptology — EUROCRYPT'96: International Conference on the Theory and Application of Cryptographic Techniques, Saragossa, Spain, May 12–16, 1996, Proceeding.* U. M. Maurer (ed.), Lecture Notes in Computer Science, Vol. 1070 (Springer), pp. 224–236.

[8] K. Kurosawa, T. Iwata, and T. Yoshiwara (2004). New covering radius of Reed–Muller codes for t-resilient functions. *IEEE Trans. Inform. Theory*, **50**(3), 468–475.

[9] C. Carlet (2020). *Boolean Functions for Cryptography and Coding Theory* (Cambridge University Press, Cambridge, England).

[10] K. Limniotis, N. Kolokotronis, and D. Kotanidis (2018). de Bruijn sequences and suffix arrays: Analysis and Constructions. In *Modern Discrete Mathematics and Analysis.* N. Daras and T. Rassias (eds.), Optimization and Its Applications, Vol. 131 (Springer Cham, Switzerland), pp. 297–316.

[11] A. J. Menezes, P. C. van Oorschot, and S. A. Vanstone (1996). *Handbook of Applied Cryptography* (CRC Press, Boca Raton, Florida).

[12] F. J. MacWilliams and N. J. A. Sloane (1983). *The Theory of Error Correcting Codes* (North-Holland Mathematical Library, Amsterdam, New York).

[13] E. R. Berlekamp (2015). *Algebraic Coding Theory*, Revised Edition (World Scientific, Singapore).

[14] J. L. Massey (1969). Shift-register synthesis and BCH decoding. *IEEE Trans. Inform. Theory*, **15**(1), 122–127.

[15] S. W. Golomb (2017). *Shift Register Sequences*, Third Revised Edition (World Scientific, Singapore).

[16] R. A. Rueppel (1986). *Analysis and Design of Stream Ciphers* (Springer, Berlin).

[17] J. L. Massey and S. Serconek (1996). Linear complexity of periodic sequences: A general theory. In *Advances in Cryptology — CRYPTO '96: 16th Annual International Cryptology Conference, Santa Barbara, California, USA, August 18–22, 1996, Proceedings*. N. Koblitz (ed.), Lecture Notes in Computer Science, Vol. 1109 (Springer), pp. 358–371.

[18] N. Kolokotronis and N. Kalouptsidis (2003). On the linear complexity of nonlinearly filtered PN-sequences. *IEEE Trans. Inform. Theory*, **49**(11), 3047–3059.

[19] N. Kolokotronis, K. Limniotis, and N. Kalouptsidis (2006). Lower bounds on sequence complexity via generalised Vandermonde determinants. In *Sequences and Their Applications — SETA 2006: 4th International Conference, Beijing, China, September 24–28, 2006, Proceedings*. G. Gong, T. Helleseth, H. Song, and K. Yang (eds.), Lecture Notes in Computer Science, Vol. 4086 (Springer), pp. 271–284.

[20] K. Limniotis, N. Kolokotronis, and N. Kalouptsidis (2006). New results on the linear complexity of binary sequences. In *Proceedings 2006 IEEE International Symposium on Information Theory, ISIT 2006, The Westin Seattle, Seattle, Washington, USA, July 9–14, 2006* (IEEE), pp. 2003–2007.

[21] N. T. Courtois and W. Meier (2003). Algebraic attacks on stream ciphers with linear feedback. In *Advances in Cryptology — EUROCRYPT 2003: International Conference on the Theory and Applications of Cryptographic Techniques, Warsaw, Poland, May 4–8, 2003, Proceedings*. E. Biham (eds.), Lecture Notes in Computer Science, Vol. 2656 (Springer), pp. 345–359.

[22] E. Dubrova and M. Hell (2017). Espresso: A stream cipher for 5G wireless communication systems. *Cryptogr. Commun.*, **9**(2), 273–289.

[23] T. Siegenthaler (1984). Correlation-immunity of nonlinear combining functions for cryptographic applications. *IEEE Trans. Inform. Theory*, **30**(5), 776–780.

[24] T. Siegenthaler (1985). Cryptanalysts representation of nonlinearly filtered ML-sequences. In *Advances in Cryptology — EUROCRYPT '85: Workshop on the Theory and Application of Cryptographic Techniques, Linz, Austria, April 1985, Proceedings*. F. Pichler (ed.), Lecture Notes in Computer Science, Vol. 219 (Springer), pp. 103–110.

[25] W. Meier and O. Staffelbach (1988). Fast correlation attacks on stream ciphers (extended abstract). In *Advances in Cryptology — EUROCRYPT'88: Workshop on the Theory and Application of Cryptographic Techniques, Davos, Switzerland, May 25–27, 1988,*

Proceedings. C. G. Günther (ed.), Lecture Notes in Computer Science, Vol. 330 (Springer), pp. 301–314.

[26] J. D. Golic (1994). Linear cryptanalysis of stream ciphers. In *Fast Software Encryption: Second International Workshop. Leuven, Belgium, December 14–16, 1994, Proceedings.* B. Preneel (ed.), Lecture Notes in Computer Science, Vol. 1008 (Springer) pp. 154–169.

[27] N. Kolokotronis, K. Limniotis, and N. Kalouptsidis (2009). Best affine and quadratic approximations of particular classes of Boolean functions. *IEEE Trans. Inform. Theory,* **55**(11), 5211–5222.

[28] N. Kolokotronis and K. Limniotis (2012). On the second-order non-linearity of cubic Maiorana–McFarland Boolean functions. In *Proceedings of the International Symposium on Information Theory and its Applications, ISITA 2012, Honolulu, HI, USA, October 28–31, 2012* (IEEE), pp. 596–600.

[29] Q. Wang, C. H. Tan, and T. F. Prabowo (2018). On the covering radius of the third order Reed–Muller code RM(3, 7). *Des. Codes Cryptogr.,* **86**(1), 151–159.

[30] Q. Wang (2019). The covering radius of the Reed–Muller code RM(2, 7) is 40, *Discrete. Math.,* **342**(12), 111625.

[31] Q. Wang and P. Stanica (2019). New bounds on the covering radius of the second order Reed–Muller code of length 128. *Cryptogr. Commun.* **11**(2), 269–277.

[32] K. Limniotis and N. Kolokotronis (2019). The error linear complexity spectrum as a cryptographic criterion of Boolean functions. *IEEE Trans. Inform. Theory,* **65**(12), 8345–8356.

[33] W. Meier, E. Pasalic, and C. Carlet (2004). Algebraic attacks and decomposition of boolean functions. In *Advances in Cryptology — EUROCRYPT 2004: International Conference on the Theory and Applications of Cryptographic Techniques, Interlaken, Switzerland, May 2–6, 2004, Proceedings.* C. Cachin and J. Camenisch (eds.), Lecture Notes in Computer Science, Vol. 3027 (Springer), pp. 474–491.

[34] C. Carlet, D. K. Dalai, K. C. Gupta, and S. Maitra (2006). Algebraic immunity for cryptographically significant Boolean functions: Analysis and construction, *IEEE Trans. Inform. Theory,* **52**(7), 3105–3121.

[35] C. Carlet (2006). On the higher order nonlinearities of algebraic immune functions. In *Advances in Cryptology — CRYPTO 2006: 26th Annual International Cryptology Conference, Santa Barbara, California, USA, August 20–24, 2006, Proceedings.* C. Dwork (ed.), Lecture Notes in Computer Science, Vol. 4117 (Springer), pp. 584–601.

[36] S. Mesnager (2008). Improving the lower bound on the higher order nonlinearity of Boolean functions with prescribed algebraic immunity. *IEEE Trans. Inform. Theory,* **54**(8), 3656–3662.

[37] P. Rizomiliotis (2010). Improving the high order nonlinearity lower bound for Boolean functions with given algebraic immunity. *Discrete Appl. Math.*, **158**(18), 2049–2055.

[38] N. T. Courtois (2003). Fast algebraic attacks on stream ciphers with linear feedback. In *Advances in Cryptology — CRYPTO 2003: 23rd Annual International Cryptology Conference, Santa Barbara, California, USA, August 17–21, 2003, Proceedings.* D. Boneh (ed.), Lecture Notes in Computer Science, Vol. 2729 (Springer), pp. 176–194.

[39] M. Liu, D. Lin, and D. Pei (2011). Fast algebraic attacks and decomposition of symmetric Boolean functions. *IEEE Trans. Inform. Theory*, **57**(7), 4817–4821.

[40] S. Mesnager and C. Tang (2021). Fast algebraic immunity of Boolean functions and LCD codes. *IEEE Trans. Inform. Theory*, **67**(7), 4828–4837.

[41] C. Carlet and K. Feng (2008). An infinite class of Balanced functions with optimal algebraic immunity, good immunity to fast algebraic attacks and good nonlinearity. In *Advances in Cryptology — ASIACRYPT 2008: 14th International Conference on the Theory and Application of Cryptology and Information Security, Melbourne, Australia, December 7–11, 2008. Proceedings.* J. Pieprzyk (ed.), Lecture Notes in Computer Science, Vol. 5350 (Springer), pp. 425–440.

[42] X. Zeng, C. Carlet, J. Shan, and L. Hu (2011). More balanced Boolean functions with optimal algebraic immunity and good nonlinearity and resistance to fast algebraic attacks. *IEEE Trans. Inform. Theory*, **57**(9), 6310–6320.

[43] K. Limniotis, N. Kolokotronis, and N. Kalouptsidis (2013). Secondary constructions of Boolean functions with maximum algebraic immunity. *Cryptogr. Commun.*, **5**(3), 179–199.

[44] J. Li, C. Carlet, X. Zeng, C. Li, L. Hu, and J. Shan (2015). Two constructions of balanced Boolean functions with optimal algebraic immunity, high nonlinearity and good behavior against fast algebraic attacks. *Des. Codes Cryptogr.*, **76**(2), 279–305.

[45] K. Limniotis and N. Kolokotronis (2018). Boolean functions with maximum algebraic immunity: Further extensions of the Carlet–Feng construction. *Des. Codes Cryptogr.*, **86**(8), 1685–1706.

[46] M. Liu, Y. Zhang, and D. Lin (2012). Perfect algebraic immune functions. In *Advances in Cryptology — ASIACRYPT 2012 — 18th International Conference on the Theory and Application of Cryptology and Information Security, Beijing, China, December 2–6, 2012. Proceedings.* X. Wang and K. Sako (eds.), Lecture Notes in Computer Science, Vol. 7658 (Springer), pp. 172–189.

[47] P. Méaux (2019). On the fast algebraic immunity of majority functions. In *Progress in Cryptology — LATINCRYPT 2019 — 6th International Conference on Cryptology and Information Security in Latin America, Santiago de Chile, Chile, October 2-4, 2019, Proceedings*, P. Schwabe and N. Thériault (eds.), Lecture Notes in Computer Science, Vol. 11774 (Springer), pp. 86–105.

[48] D. Tang (2020). A note on the fast algebraic immunity and its consequences on modified majority functions. *Adv. Math. Commun.*, **14**(1), 111–125.

[49] P. Méaux (2021). On the fast algebraic immunity of threshold functions. *Cryptogr. Commun.*, **13**(5), 741–762.

[50] K. Limniotis (2013). Algebraic attacks on stream ciphers: Recent developments and new results. *J. Appl. Math. & Bioinf.*, **3**(1), 57–81.

[51] N. Kalouptsidis and K. Limniotis (2004). Nonlinear span, minimal realizations of sequences over finite fields and de Bruijn generators. In *Proceedings of the International Symposium on Information Theory and its Applications, ISITA 2004, Parma, Italy, October 10–13, 2004*. IEICE, Tokyo, Japan, pp. 794–799.

[52] A. Lempel (1970). On a homomorphism of the de Bruijn graph and its applications to the design of feedback shift registers. *IEEE Trans. Comput.*, **19**(12), 1204–1209.

[53] J. Mykkeltveit, M. Siu, and P. Tong (1979). On the cycle structure of some nonlinear shift register sequences. *Inform. Control.*, **43**(2), 202–215.

[54] R. A. Games (1983). A generalized recursive construction for de Bruijn sequences. *IEEE Trans. Inform. Theory*, **29**(6), 843–849.

[55] K. Mandal and G. Gong (2012). Cryptographically strong de Bruijn sequences with large periods. In *Selected Areas in Cryptography: 19th International Conference, SAC 2012, Windsor, ON, Canada, August 15–16, 2012, Revised Selected Papers*, L. R. Knudsen and H. Wu (eds.), Lecture Notes in Computer Science, Vol. 7707 (Springer), pp. 104–118.

[56] C. J. A. Jansen, W. G. Franx, and D. E. Boekee (1991). An efficient algorithm for the generation of de Bruijn cycles. *IEEE Trans. Inform. Theory.*, **37**(5), 1475–1478.

[57] J. Mykkeltveit and J. Szmidt (2013). On cross joining de Bruijn sequences, *IACR Cryptology ePrint Archieve* p. 760. http://eprint.iacr.org/2013/760.

[58] T. Helleseth and T. Kløve (1991). The number of cross-join pairs in maximum length linear sequences. *IEEE Trans. Inform. Theory*, **37**(6), 1731–1733.

[59] T. Helleseth (2005). de Bruijn sequence. In *Encyclopedia of Cryptography and Security*. H. C. A. van Tilborg (ed.), (Springer, Cham, Switzerland), pp. 138–140.

[60] E. Dubrova (2014). Generation of full cycles by a composition of NLFSRs. *Des. Codes Cryptogr.*, **73**(2), 469–486.

[61] K. Mandal and G. Gong (2016). Feedback reconstruction and implementations of pseudorandom number generators from composited de Bruijn sequences. *IEEE Trans. Comput.*, **65**(9), 2725–2738.

[62] B. Yang, K. Mandal, M. D. Aagaard, and G. Gong (2017). Efficient composited de Bruijn sequence generators. *IEEE Trans. Comput.*, **66**(8), 1354–1368.

[63] J. Dong and D. Pei (2017). Construction for de Bruijn sequences with large stage. *Des. Codes Cryptogr.*, **85**(2), 343–358.

[64] A. H. Chan, R. A. Games, and E. L. Key (1982). On the complexities of de Bruijn sequences. *J. Combin. Theory Ser. A*, **33**(3), 233–246.

[65] T. Etzion and A. Lempel (1984). Construction of de Bruijn sequences of minimal complexity. *IEEE Trans. Inform. Theory*, **30**(5), 705–708.

[66] T. Etzion (1999). Linear complexity of de Bruijn sequences — Old and new results. *IEEE Trans. Inform. Theory*, **45**(2), 693–698.

[67] M. S. Turan (2012). On the nonlinearity of maximum-length NFSR feedbacks. *Cryptogr. Commun.*, **4**(3–4), 233–243.

[68] U. Manber and E. W. Myers (1993). Suffix arrays: A new method for on-line string searches. *SIAM J. Comput.*, **22**(5), 935–948.

[69] D. K. Kim, J. S. Sim, H. Park, and K. Park (2005). Constructing suffix arrays in linear time. *J. Discrete Algorithms*, **3**(2–4), 126–142.

Index